T0339856

Colloidal Gold Nanorods

This book covers the synthesis and application of colloidal gold nanorods including their properties, approaches for various chemical synthesis, and different gold nanorod-based nanocomposites with their properties and application potentials. Furthermore, it covers the surface chemistry and functionalization of gold nanorods for numerous biomedical applications. Various applications of gold nanorods including optical probes, dark filed contrast agents, photothermal therapy agents, and plasmonic photocatalysts are covered, along with the toxicological aspects.

Features:

- Covers all aspects of gold nanorods along with selected protocols.
- Focuses on synthetic chemistry, optical property, and functionalization approach of colloidal gold nanorods.
- Describes standard synthetic methods and advantages of gold nanorods in biomedical applications.
- Includes authentic and reproducible experimental procedures.
- Discusses applications like redox catalysts, catalyst promoters, delivery carriers, solar cell materials, and so forth.

This book is aimed at graduate students and researchers interested in nanotechnology and gold nanoparticles.

Nikhil Ranjan Jana is a senior professor at the School of Materials Science, Indian Association for the Cultivation of Science, Kolkata, India. He earned his undergraduate degree (1987) from Midnapore College, India, and his masters (1989) and doctoral (1994) degrees from the Indian Institute of Technology, Kharagpur, India. Prof. Jana worked as a postdoctoral fellow at the University of South Carolina, USA (1999–2001) and the University of Arkansas, USA (2003) and as a scientist at the Institute of Bioengineering and Nanotechnology, Singapore (2004–2008). His research group designed colloidal nanobioconjugates for controlling cellular processes and developed nanoprobes/nanodrugs for sub-cellular targeting/imaging and inhibiting amyloid aggregation under intra-/extra-cellular space. His group published 180 peer-reviewed research articles, which have received approximately 25,000 citations, in internationally recognized journals.

Colloidal Gold Nanorods

Science and Technology

Nikhil Ranjan Jana

CRC Press
Taylor & Francis Group
Boca Raton London New York

CRC Press is an imprint of the
Taylor & Francis Group, an **informa** business

First edition published 2023
by CRC Press
6000 Broken Sound Parkway NW, Suite 300, Boca Raton, FL 33487-2742

and by CRC Press
4 Park Square, Milton Park, Abingdon, Oxon, OX14 4RN
CRC Press is an imprint of Taylor & Francis Group, LLC

ISBN: 9781032156835 (hbk)
ISBN: 9781032156842 (pbk)
ISBN: 9781003245339 (ebk)

DOI: 10.1201/9781003245339

Typeset in Times
by Deanta Global Publishing Services, Chennai, India

Contents

Preface

The goal of writing this book is to cover all aspects of gold nanorods in a single volume. Considering the interdisciplinary nature of the field, extensive amount of scientific literature on this subject, and involvement of a wide variety of researchers from different scientific backgrounds, a single book covering these aspects is important for beginners in this field.

This book consists of 12 chapters: properties of gold nanorod, chemical synthesis of colloidal gold nanorod, gold nanorod-based different nanocomposites, surface chemistry and functionalization of gold nanorod, plasmonic property of gold nanorod for optical probe, gold nanorod as imaging contrast agent, application of gold nanorod in photothermal therapy, gold nanorod as plasmonic photocatalyst, gold nanorod in electrochemical applications, other applications of gold nanorod, toxicology of gold nanorod, and advantages of gold nanorod.

Most of the content of this book is drawn from different published articles from various research groups. I am very thankful to those publishers and authors from whose publications I have collected useful information. I am also thankful to some pioneers of related fields with whom I have interacted in various ways and occasions during my long research career. I would specifically like to mention Prof. J. Allan Creighton, Prof. Henk Lekkerkerker, Prof. Mostafa El-Sayed, Prof. Martin Moskovits, Prof. Paul Alivisatos, and Prof. Jim Heath. I may not have cited their works due to limited space and scope, however, their research and publications have strongly influenced my research and inspired me in writing this book.

I hope this book may provide guidance to graduates and researchers, particularly those who are beginning their careers in the field of experimental nanoscience. This book could also be used in nanobiotechnology courses.

Professor Nikhil Ranjan Jana
Indian Association for the Cultivation of Science
Kolkata, India

Acknowledgments

I like to thank Prof. Tarasankar Pal, Prof. Catherine J. Murphy, Prof. Xiaogang Peng, and Prof. Jackie Y. Ying with whom I have had a long association during my research career. I initially learned about experimental nanoscience in the laboratory of Prof. Pal at IIT, Kharagpur, explored all possible research ideas with a great deal of freedom in the laboratory of Prof. Murphy at the University of South Carolina, Columbia, increased my understanding of nanoparticle synthetics in the laboratory of Prof. Peng at the University of Arkansas, and became an independent researcher in the field of nanobiotechnology under the challenging leadership of Prof. Jackie Ying at the Institute of Nanoscience and Nanotechnology, Singapore. I would like to convey my deep and sincere respect to everyone who helped me to shape my research career. I would also like to thank my research group members who are involved in developing this field of research and have supported me in writing this book.

Properties of Gold Nanorod

1

1.1 INTRODUCTION

The fascinating color of finely divided gold particles has been appreciated since ancient times.[1,2] Colloidal gold has been used in ancient times as a method of staining glass. Modern scientific evaluation of colloidal gold began with Michael Faraday's work in the 1850s[1,2] (Figure 1.1). He prepared the first pure sample of colloidal gold in 1857, and it is still optically active. Faraday recognized that the color of the colloidal gold was caused by the small size of the gold particles. In 1898, Richard Adolf Zsigmondy prepared the first colloidal gold in dilute solution, and in 1908 Gustav Mie provided the theory for light scattering that describes the optical properties of colloidal gold particles.[1,2]

With the advances in various analytical technologies in the 20th century, studies on gold nanoparticles have accelerated. Advanced microscopy methods, such as electron microscopy, have contributed the most to nanoparticle research. This is particularly because small particles can be directly observed under electron microscopes and their quality in terms of size and size distribution can be evaluated. Such microscopic study shows that the citrate reduction-based synthesis of gold nanoparticles, pioneered by J. Turkevich et al. in 1951 and refined by G. Frens in the 1970s, produces modestly monodispersed spherical gold nanoparticles of around 10–20 nm in diameter.[3,4]

The stability of the colloidal dispersion of gold nanoparticles without any precipitation is a critical issue related to the study of their properties. However, most of the colloidal golds are stable in dilute solution. Although such dispersion can be used to study various physical properties, it is difficult to use these dispersions for large-scale synthesis and further processing of gold nanoparticles. The discovery of the Brust synthetic method in 1994 opened the door for large-scale synthesis and processing of gold nanoparticles.[5] This method produces gold nanoparticles in a water-toluene two-phase mixture that are 2--5 nm in size, and the nanoparticles are capped with monolayers of thiol-based surfactants (e.g. dodecanethiol). The size distribution of these nanoparticles can be narrowed by heating the colloidal solution from 80 to 100 °C.[6] The most

DOI: 10.1201/9781003245339-1

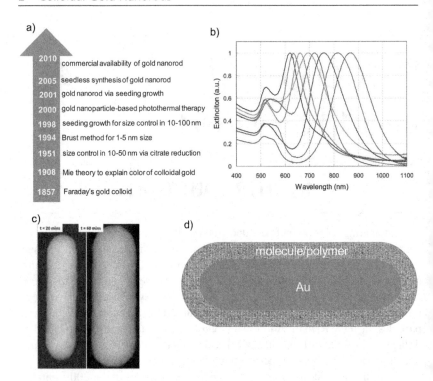

FIGURE 1.1 a) Timeline of important developments in the science and technology of gold nanorods. b) Normalized absorption spectra of colloidal gold nanorods of different lengths with different longitudinal plasmon resonance bands. Reprinted with permission from Hu, R. et al. 2009. Metallic nanostructures as localized plasmon resonance enhanced scattering probes for multiplex dark-field targeted imaging of cancer cells. *The Journal Of Physical Chemistry C*, 113, 2676–2684. Copyright 2009 American Chemical Society. c) High-angle annular dark-field scanning transmission electron microscopy images of gold nanorods at two different stages of growth. Reprinted with permission from Walsh, M. J. et al. 2017. A mechanism for symmetry breaking and shape control in single crystal gold nanorods. *Accounts of Chemical Research*, 50, 2925–2935. Copyright 2017. American Chemical Society. d) Typical core-shell structure of commonly used colloidal gold nanorods.

important advantage of this approach is that nanoparticles can be made on a gram scale, are dispersible in the organic/aqueous phase, and can be handled like a chemical reagent. We can use nanoparticles as a chemical, dissolve them in a solvent, precipitate them by adding another solvent, use them to react with other chemicals, link them with other chemicals/biochemicals, and exploit them as a reagent.

The understanding of the nucleation-growth mechanism has further advanced the size-controlled synthesis of gold nanoparticles. It has been observed that nucleation is associated with a high energy barrier and nucleation centers can catalyze particle growth as well as secondary nucleation.[7] Thus identifying a growth condition that restricts any secondary nucleation becomes important and appears very challenging.[8] In this context, a seed-mediated growth approach has been explored and developed where preformed smaller gold nanoparticles are used as nucleation centers to grow the nanoparticles into a larger size.[9,10]

Generally, the use of synthetic methods produces mainly spherical gold nanoparticles, and this shape is expected as it has the lowest surface area with the lowest surface energy. However, nanoparticles with anisotropic shapes are observed to form occasionally depending on the growth condition and stabilizers used.[8–10] Thus, in those days synthetic approaches for anisotropic gold nanorods relied on alternative methods that utilized various porous templates.[11] The extensive utilization of the seed-mediated growth method hinted that gold nanorods might be produced in high yield without using any template. It was observed that this approach enhances the yield of anisotropic nanoparticles, particularly nanorods.[8–10] The enhanced nanorod yield was primarily due to the minimization of secondary nucleation and overdriven growth kinetics. With this understanding and advancements in technology, the seeding growth of gold nanorods with high yield was developed in 2001.[12,13] This method is advantageous because gold nanorods can be synthesized in test tubes through simple mixing of chemicals. After this stage, a variety of modifications were made to this method, and it was reported that the modifications improved the quality of nanorods.[14] Most importantly, a seedless method for gold nanorod growth was developed in 2005.[15] These discoveries and methods have boosted the field significantly and gold nanorods of different lengths and surface chemistry are now commercially available.

The field of gold nanorod-based application is well advanced and mature now. This is particularly because of the commercial availability of gold nanorods and the development of a wide variety of chemical approaches for their synthesis, surface modification, and chemical processing. In addition, the application potential of gold nanorods to all possible areas has been well explored and advanced. Table 1.1 summarizes the important properties of gold nanorods that are utilized in different applications. These properties include nanometer size, colloidal property, shape-dependent plasmonic property, photo-responsive property, and self-assembly property. There are several good reviews on gold nanorods that cover certain select areas.[14,16–23] In particular, these reviews cover synthetic advancements,[14,16,17] biomedical application,[18,21] plasmonic properties of supercrystals,[19,20] and analytical applications.[22] Figure 1.1 highlights some of the key properties of gold nanorods.[24,25]

TABLE 1.1 Summary of Common Properties of Au Nanorods That Are Important in Basic and Applied Science.

PROPERTIES	DETAILS OF THE PROPERTY	WHY IS THIS PROPERTY IMPORTANT?
Primary chemical composition	gold, hexadecyltrimethylammonium bromide (CTAB)	gold offers good chemical stability, adsorbed CTAB offers dispersibility in water
Size	6–15 nm diameter, 10–200 nm length	virus-like size, offers colloidal property and endocytic uptake into cell
Colloidal property	dispersible in water without precipitation	easier processing, accessibility to biological interface
Optical property	absorb and scatter light in the visible to NIR region	plasmon-based detection and imaging
Photo-responsive property	Light-induced local heating	photothermal therapy, light-induced delivery
Surface property	charged (either cationic or anionic), terminated with defined chemicals/biochemicals	selective biological labeling
Self-assembly	entropy-driven self-assembly to liquid crystalline structure at high concentration	processing and device fabrication

In this book, we will cover all of these aspects in 12 different chapters. Chapter 2 focuses on the history of the synthetic development of colloidal gold nanorods and discusses synthetic advancements in making the high-quality nanorods that are currently used. Chapter 3 focuses on gold nanorod-based different nanocomposites that are designed to enhance the application potential of gold nanorods. Chapter 4 focuses on the coating chemistry and functionalization of gold nanorods that are essential for their application. Chapter 5 focuses on the application of gold nanorods as optical probes for detecting various chemicals and biochemicals. Chapter 6 summarizes the application of gold nanorods as contrast agents in different types of imaging. Chapter 7 summarizes the application of gold nanorods in photothermal therapy. Chapter 8 summarizes the application of gold nanorods as plasmonic photocatalysts. The application of gold nanorods in various electrochemical reactions is summarized in Chapter 9. Chapter 10 summarizes other unique applications of gold nanorods, including their use as drug delivery carriers, in remote-controlled drug delivery, and in light-responsive targeting. Chapter 11 summarizes the toxicological aspects of gold nanorods. Chapter 12 discusses the advantages of gold nanorods over other gold nanoparticles.

1.2 CHEMICAL COMPOSITION OF GOLD NANOROD

The primary components of gold nanorods are metallic gold and surfactant stabilizer present around the nanorod.[12–16] In addition, silver ions may be present in small amounts, as they are used during synthesis. Silver ions are adsorbed at the nanorod's surface in the form of salt or ions associated with the surfactant stabilizer. The main surfactant component present is hexadecyltrimethylammonium bromide (CTAB) which is used as a shape-directing agent during synthesis. CTAB is present as a capping agent after the nanorod is purified. Other surfactants are sometimes used during synthesis and may also be present in small amounts. These include hexadecyltrimethylammonium chloride (CTAC), benzyldimethylhexadecylammonium chloride (BDAC), and tetraoctylammonium bromide (TOAB). The described compositions of metals and surfactants are restricted to synthesized nanorods. However, the composition can vary after surface modification with other chemicals or composite formation with other organic/inorganic materials. These compositions are discussed in detail in Chapters 2 and 3.

The presence of gold offers good chemical stability, as gold is one of the noble metals and has low chemical reactivity. However, gold's nanosize and anisotropic shape increases its chemical reactivity towards some of the well-known chemicals such as cyanide-oxygen and peroxide.[26] The adsorbed CTAB creates the colloidal property of gold nanorods by charging the surface with water-soluble bi-layer structures.[14]

1.3 SIZE OF GOLD NANOROD

One important aspect of gold nanorods is their size in nanometers. The overall size of gold nanorods includes their inorganic core and molecular/polymeric shell. The typical length of the inorganic core of a gold nanorod varies from

10 to 200 nm and the diameter varies between 6 and 15 nm.[23] However, the length can be longer, in which case, the nanorod is called a nanowire. Shell materials are usually a few nanometers thick and thus increase the overall size by another ~10 nm. While the inorganic core is critical for showing plasmonic properties, the organic/inorganic shell is important for colloidal properties and responsible for interactions with the biological interface. The overall size of a nanorod is similar to the typical size of viruses, biopolymers, and biological components.[21] Combined with specific surface chemistry, this size is ideal for endocytic entry into living cells and trafficking to different subcellular compartments. An anisotropic shape has the unique option for entry into cells either via longitudinal or transverse modes.[27] The overall nanometer dimension and shape anisotropy found in nanorods offer unique opportunities to explore their various physical properties and applications.

1.4 COLLOIDAL PROPERTY OF GOLD NANOROD

Another important property of gold nanorods is that they are colloidal in nature.[21,23] In most cases they are dispersible in water without any precipitation. However, in certain selected cases, they can be dispersed in the organic phase after suitable surface modification. The colloidal nature of gold nanorods arises due to three specific reasons. First, they are small in size and Brownian motion allows their mobility in solvents against the precipitation under gravitational attraction. Second, they have charged surfaces that arise due to adsorbed ions/molecules/polymers. Such charged surfaces restrict the particle–particle interaction and agglomeration. Third, surface chemistries are designed for high solubility in the dispersed phase. This allows their dispersion stability.

The colloidal property of gold nanorods has three unique advantages. First, the brilliant color of gold nanorods can be observed at their colloidal dispersion stage. In addition, the colloidal solution can be exposed using light and absorbed or scattered light can be detected under appropriate instrumentation. Second, the colloidal state offers easier processing. In particular, it is useful for surface chemistry exercises and can be transformed into film or composites. Third, colloidal forms are accessible to various biological interfaces. In particular, colloidal nanorods can travel from injection sites to different organs, tissue components, and subcellular compartments of a living animal. Overall colloidal forms allow a wider application of gold nanorods.

1.5 OPTICAL PROPERTY OF GOLD NANOROD

Gold nanorods are interesting due to their unique optical properties.[18,19] In particular, their colloidal solutions display an intense plasmonic absorption band that can be tuned from a visible to near-infrared (NIR) spectrum by changing the length-to-width ratio. In addition, they can scatter light depending on their size/length and wavelengths can be increased by increasing the size/length of gold nanorods. This plasmonic property depends on the dispersed state, dispersed medium, adsorbed surface molecules, and extent of particle aggregation. This plasmonic property has been utilized for a wide variety of optical-based detection applications. In addition, the light scattering property can be used for the application of dark field imaging-based detection. Compared to molecular probes and dyes, these optical properties have three unique advantages. First, the color is more intense and similar to well-known dyes with high molar extinction coefficients. Second, the color is very stable under exposure to continuous light without any photo-bleaching issues. This provides an advantage over the application of molecular probes for imaging or detection. Third, similar surface chemistry can be adapted to nanorods of different lengths and thus optical probes with different lengths of nanorods can be easily designed. In contrast, different synthetic chemistry is necessary to design different molecular probes.

1.6 PHOTO-RESPONSIVE PROPERTY

Gold nanorods have a unique plasmonic property that offers additional features to manipulate light for chemical and physical processes.[18,23] In particular, colloidal gold nanorods can be used to capture light from visible to NIR ranges by varying the length of the nanorod. The captured light can be used for three different chemical/physical processes. First, photochemical reactions can be performed at the nanorod interface, which is commonly known as plasmonic photocatalysis. Second, nanorods can be used to capture sunlight and convert it into chemical and electrical energy. Third, captured light offers photothermal effects that result in an increase in the temperature of the surrounding medium which can be tens of degrees. This photothermal effect can be used for photothermal therapy and light-induced drug delivery applications.

1.7 ASSEMBLY PROPERTY OF GOLD NANOROD

Gold nanorods have an anisotropic property that offers unique assembly properties. In particular, they self-assemble into liquid crystalline structures at high concentrations.[28–30] For example, smaller nanorods can be used to create smectic structures, longer nanorods assemble into columnar or ribbon-like structures, and long nanorods and nanowires make bundle structures. However, in the case of nanorods with wider length distribution, nematic structures are the most common. There are two accepted reasons for such self-assembly types. First, at high concentrations nanorods have restricted Brownian motion due to space limitations and under this circumstance, they maximize their rotational freedom via liquid crystal-like ordering. Entropy is the primary driving force for such self-assembly. Second, capped surfactant at the nanorod surface induces side-by-side interaction between nanorods via hydrophobic surface tails.

In addition, small molecules and polymers are used to assemble nanorods into different structures.[16–20] In particular, thiol-based small molecules/polymers are used to make linear chains of nanorods via end-to-end assembly. Similarly, Langmuir–Blodgett films, drying-induced assembly, and other approaches are used to make two dimensional (2D) and three dimensional (3D) gold nanorod structures. These assembly properties are used for shape separation and nanorod processing and to fabricate devices that are subsequently used for optical detection and solar energy conversion.

1.8 COMPARISON WITH OTHER ANISOTROPIC NANOPARTICLES

There are several other anisotropic nanoparticles with properties that are somewhat similar to gold nanorods. Among them, gold and silver-based anisotropic nanoparticles are the most important in terms of optical properties. Gold-based nanoparticles include gold platelets,[31] triangular gold nanoprisms,[32] gold nanocubes,[33] gold nanostars,[34] and gold nanoshells.[35] Although their plasmonic properties can be tuned from visible to NIR spectrums, such tunability is either restricted to narrow wavelength regions or their plasmon bands are relatively broad. Silver-based anisotropic nanoparticles include nanorods, platelets, nanoprisms, and nanostars.[36,37] These particles have relatively more intense plasmon bands, but they are less biocompatible, chemically less stable, and more susceptible to shape change under adverse conditions.

Other important non-metallic anisotropic nanoparticles include CdSe nanorods,[38] TiO_2 nanorods,[39] iron oxide nanocubes,[40] hydroxyapatite nanorods/nanowires,[41] and $BaTiO_3$ nanorods.[42] Among them CdSe-based nanorods have fluorescent properties, TiO_2 nanorods are used in photocatalysis, iron oxide nanocubes have magnetic properties, hydroxyapatite nanorods/nanowires have biomedical application potential, and $BaTiO_3$ nanorods are used as piezocatalytic materials. All of these materials are colloidal in nature, a property that makes them similar to colloidal gold nanorods.

1.9 CONCLUSION

This chapter discussed the historical background and development of the field of gold nanorods and their general properties. This includes the use of gold nanoparticles in ancient times, modern scientific evaluation of colloidal gold, large-scale synthesis, processing of gold nanoparticles, and finally the development of gold nanorod synthesis method in 2001. Next, we briefly discussed the general properties of gold nanorods that are important for their application. We also indicated the different applications of gold nanorods that are covered in various chapters of this book.

REFERENCES

1. Daniel, M.-C. and Astruc, A. 2004. Gold nanoparticles: Assembly, supramolecular chemistry, quantum-size-related properties, and applications toward biology, catalysis, and nanotechnology. *Chemical Reviews*, 104, 293–346.
2. Saha, K., Agasti, S. S., Kim, C., Li, X. and Rotello, V. M. 2012. Gold nanoparticles in chemical and biological sensing. *Chemical Reviews*. 112, 2739–2779.
3. Turkevich, J., Stevenson, P. C. and Hillier, J. 1951. A study of the nucleation and growth processes in the synthesis of colloidal gold. *Discussions of the Faraday Society*, 11, 55–75.
4. Frens, G. 1973. Controlled nucleation for the regulation of the particle size in monodisperse gold suspensions. *Nature Physical Science*, 241, 20–22.
5. Brust, M., Walker, M., Bethell, D., Schiffrin, D. J. and Whyman, R. 1994. Synthesis of thiol-derivatised gold nanoparticles in a two-phase liquid–liquid system. *Journal of the Chemical Society, Chemical Communications*, 801–802.

6. Shimpi, J. R., Sidhaye, D. S. and Prasad, B. L. V. 2017. Digestive ripening: A fine chemical machining process on the nanoscale. *Langmuir*, 33, 9491–9507.

7. Watzky, M. A. and Finke, R. G. 1997. Transition metal nanocluster formation kinetic and mechanistic studies. A new mechanism when hydrogen is the reductant: Slow, continuous nucleation and fast autocatalytic surface growth. *Journal of The American Chemical Society*, 119, 10382–10400.

8. Jana, N. R., Gearheart, L. and Murphy, C. J. 2001. Evidence for seed-mediated nucleation in the chemical reduction of gold salts to gold nanoparticles, *Chemistry of Materials*, 13, 2313–2322.

9. Brown, K. R., Walter, D. G. and Natan, M. J. 2000. Seeding of colloidal Au nanoparticle solutions. 2. Improved control of particle size and shape. *Chemistry of Materials*, 12, 306–313.

10. Jana, N. R., Gearheart, L. and Murphy, C. J. 2001. Seeding growth for size control of 5–40 nm diameter gold nanoparticles. *Langmuir*, 17, 6782–6786.

11. van der Zande, B. M. I., Bo¨hmer, M. R., Fokkink, L. G. and Schönenberger, C. 2000. Colloidal dispersions of gold rods: Synthesis and optical properties. *Langmuir*, 16, 451–458.

12. Jana, N. R., Gearheart, L. and Murphy, C. J. 2001. Wet chemical synthesis of high aspect ratio cylindrical gold nanorods. *The Journal of Physical Chemistry B*, 105, 4065–4067.

13. Jana, N. R., Gearheart, L. and Murphy, C. J. 2001. Seed-mediated growth approach for shape-controlled synthesis of spheroidal and rod-like gold nanoparticles using a surfactant template. *Advanced Materials*, 13, 1389–1393.

14. Lohse, S. E. and Murphy, C. J. 2013. The quest for shape control: A history of gold nanorod synthesis. *Chemistry of Materials*, 25, 1250–1261.

15. Jana, N. R. 2005. Gram-scale synthesis of soluble, near-monodisperse gold nanorods and other anisotropic nanoparticles. *Small*, 1, 875–882.

16. Perez-Juste´, J., Pastoriza-Santos, I., Liz-Marzan´, L. M. and Mulvaney, P. 2005. Gold nanorods: Synthesis, characterization and applications. *Coordination Chemistry Reviews*, 249, 1870–1901.

17. Sharma, V., Park, K. and Srinivasarao, M. 2009. Colloidal dispersion of gold nanorods: Historical background, optical properties, seed-mediated synthesis, shape separation and self-assembly. *Materials Science and Engineering: R: Reports*, 65, 1–38.

18. Huang, X., Neretina, S. and El-Sayed, M. A. 2009. Gold nanorods: From synthesis and properties to biological and biomedical applications. *Advanced Materials*, 21, 4880–4910.

19. Chen, H., Shao, L., Lia, Q. and Wang, J. 2013. Gold nanorods and their plasmonic properties. *Chemical Society Reviews*, 42, 2679–2724.

20. Scarabelli, L., Hamon, C. and Liz-Marzan, L. M. 2017. Design and fabrication of plasmonic nanomaterials based on gold nanorod supercrystals. *Chemistry of Materials*, 29, 15–25.

21. Murphy, C. J., Chang, H.-H., Falagan-Lotsch, P., Gole, M. T., Hofmann, D. M., Hoang, K. N. L., McClain, S. M., Meyer, S. M., Turner, J. G., Unnikrishnan, M., Wu, M., Zhang, X. and Zhang, Y. 2019. Virus-sized gold nanorods: Plasmonic particles for biology. *Accounts for Chemical Research*, 52, 2124–2135.

22. Gorbunova, M., Apyari, V., Dmitrienko, S. and Zolotov, Y. 2020. Gold nanorods and their nanocomposites: Synthesis and recent applications in analytical chemistry. *Trends in Analytical Chemistry*, 130, 115974.

23. Zheng, J., Cheng, X., Zhang, H., Bai, X., Ai, R., Shao, L. and Wang, J. 2021. Gold nanorods: The most versatile plasmonic nanoparticles. *Chemical Reviews*, 121, 13342–13453.

24. Hu, R., Yong,K.-T., Roy, I., Ding, H., He, S. and Prasad, P. N. 2009. Metallic nanostructures as localized plasmon resonance enhanced scattering probes for multiplex dark-field targeted imaging of cancer cells. *The Journal of Physical Chemistry C*, 113, 2676–2684.

25. Walsh, M. J., Tong, W., Katz-Boon, H., Mulvaney, P., Etheridge, J. and Funston, A. M. 2017. A mechanism for symmetry breaking and shape control in single crystal gold nanorods. *Accounts of Chemical Research*, 50, 2925–2935.

26. Jana, N. R., Gearheart, L., Obare, S. O. and Murphy, C. J. 2002. Anisotropic chemical reactivity of gold spheroids and nanorods, *Langmuir*, 18, 3, 922–927.

27. Dasgupta, S., Auth, T. and Gompper, G. 2014. Shape and orientation matter for the cellular uptake of nonspherical particles. *Nano Letters*, 14, 687–693.

28. Jana, N. R. 2004. Shape effect in nanoparticle self-assembly. *Angewandte Chemie International Edition*, 43, 1536–1540.

29. Jana, N. R. 2003. Nanorod shape separation using surfactant assisted self-assembly. *Chemicals Communications*, 1950–1951.

30. Vaia, R. A. 2010. Depletion-induced shape and size selection of gold nanoparticles. *Nano Letters*, 10, 1433–1439.

31. Chen, L., Ji, F., Xu, Y., He, L., Mi, Y., Bao, F., Sun, B., Zhang, X. and Zhang, Q. 2014. High-yield seedless synthesis of triangular gold nanoplates through oxidative etching. *Nano Letters*, 14, 7201–7206.

32. Langille, M. R., Personick, M. L., Zhang, J. and Mirkin, C. A. 2012. Defining rules for the shape evolution of gold nanoparticles. *Journal of the American Chemical Society*, 134, 14542–14554.

33. Park, J.-E., Lee, Y. and Nam, J.-M. 2018. Precisely shaped, uniformly formed gold nanocubes with ultrahigh reproducibility in single-particle scattering and surface-enhanced Raman scattering. *Nano Letters*, 18, 6475–6482.

34. Sau, T. K., Rogach, A. L., Döblinger, M. and Feldmann, J. 2011. One-step high-yield aqueous synthesis of size-tunable multispiked gold nanoparticles. *Small*, 7, 2188–2194.

35. Brinson, B. E., Lassiter, J. B., Levin, C. S., Bardhan, R., Mirin, N. and Halas, N. J. 2008. Nanoshells made easy: Improving Au layer growth on nanoparticle surfaces. *Langmuir*, 24, 14166–14171.

36. Jana, N. R. and Pal, T. 2007. Anisotropic metal nanoparticles for use as surface-enhanced Raman substrates. *Advanced Materials*, 19, 1761–1765.

37. Zhang, O., Li, N., Goebl, J., Lu, Z. and Yin, Y. 2011. A systematic study of the synthesis of silver nanoplates: Is citrate a "magic" reagent? *Journal of the American Chemical Society*, 133, 18931–18939.

38. Peng, X., Manna, L., Yang, W., Wickham, J., Scher, E., Kadavanich, A. and Alivisatos, A. P. 2000. Shape control of CdSe nanocrystals. *Nature*, 404, 59–61.

39. Mukherjee, K., Acharya, K., Biswas, A. and Jana, N. R. 2020. TiO_2 nanoparticles co-doped with nitrogen and fluorine as visible-light-activated antifungal agents. *ACS Applied Nano Materials*, 3, 2016–2025.

40. Wang, O., Ma, X., Liao, H., Liang, Z., Li, F., Tian, J. and Ling, D. 2020. Artificially engineered cubic iron oxide nanoparticle as a high-performance magnetic particle imaging tracer for stem cell tracking. *ACS Nano*, 14, 2053–2062.

41. Das, P. and Jana, N. R. 2016. Length-controlled synthesis of calcium phosphate nanorod and nanowire and application in intracellular protein delivery. *ACS Applied Materials and Interfaces*, 8, 8710–8720.

42. Biswas, A., Saha, S., Pal, S. and Jana, N. R. 2020. TiO_2-templated $BaTiO_3$ nanorod as a piezocatalyst for generating wireless cellular stress. *ACS Applied Materials and Interfaces*, 12, 48363–48370.

Chemical Synthesis of Colloidal Gold Nanorod

2

2.1 INTRODUCTION

Gold nanorods show shape dependent optical phenomena that are not seen in spherical gold nanoparticles.[1-7] These colloidal gold nanorods exhibit absorption bands that are tunable from a visible to NIR spectrum. This arises due to the emergence of transverse plasmon bands (linked to short axis or diameter) and more importantly longitudinal plasmon bands (linked to long axis or length). These plasmonic bands are sensitive to local chemical/physical environments, and they are used for sensing, plasmon-enhanced spectroscopies, biomedical imaging, and photothermal therapy applications.

However, the synthesis of anisotropic gold nanorods is more challenging than spherical gold nanoparticles.[8-34] This is particularly because a spherical shape has minimal surface area (hence lowest surface energy) and the highly symmetric cubic crystal structure of gold tries to attain pseudo-spherical shapes to minimize surface energy. Thus early approaches for gold nanorod synthesis primarily relied on hard template-based approaches where the pores and channels of a template dictated the shape anisotropy.[9] These approaches have limitations in making high-quality nanorods with large-scale synthesis options. Synthetic breakthrough occured with the advent of the discovery of colloid-chemical approaches during 1997 to 2003, along with step-by-step improvements of the methods in later stages (see Table 2.1 and Figures 2.1–2.4). High-quality gold nanorods were first prepared in 1997 via an electro-chemical approach at the electrode surface[8] and then via a seeding growth approach in 2001 in a test tube.[10,11] The seeding growth approach was further

DOI: 10.1201/9781003245339-2

TABLE 2.1 History of Au Nanorod Synthetic Approaches That Are Gradually Developed and the Primary Shape Inducing Conditions/Chemicals.

SYNTHETIC APPROACH	PRODUCT TYPE (PLASMONIC ABSORPTION)	LENGTH, WIDTH, ASPECT RATIO	PRIMARY SHAPE DIRECTING AGENT/CONDITION#	REFERENCE (YEAR)
Electrochemical at electrode surface	nanorod (600–1100 nm)	20–150 nm, 10–15 nm, 2–11	CTAB, TOAB, Ag+	8 (1997)
Electrochemical in a hard template	nanorod (650–1700 nm)	40–700 nm, 12–22 nm, 2–50	nanopores of alumina	9 (2000)
First report via colloid-chemical seeding growth	nanorod, spheroid (600–1800 nm)	50–300 nm, 10–16 nm, 5–18	3.5 nm seed, CTAB, Ag+	10,11 (2001)
Photochemical	nanorod (600–800 nm)	40–80 nm, 12–18 nm, 3–5	CTAB, Ag+	12 (2002)
First report of nanorod shape separation via depletion attraction	long nanorod (1400–1800 nm)	200–300 nm, 12–18 nm, 18–20	3.5 nm seed, CTAB	46 (2003)
Colloidal seeding growth of Au sphere	nanorod (600–1300 nm)	30–100 nm, 6–12 nm, 1.5–10	1.5 nm seed, CTAB, BDAC, Ag+	13 (2003)
Colloidal seeding growth of Au sphere	nanorod, rectangle, cube, triangle, star (600–1000 nm)	70–300 nm, 20–50 nm, 1–5	CTAB, Ag+	14,15 (2004)
First report of colloid-chemical seedless approach	nanorod (600–1000 nm)	10–50 nm, 4–10 nm, 2–5	CTAB, Ag+	17 (2005)
Oriented attachment of Au spheres	nanowire (600–1000 nm)	micron, 2 nm	oleic acid, olylamine	18 (2007)

(Continued)

TABLE 2.1 (CONTINUED) History of Au Nanorod Synthetic Approaches That Are Gradually Developed and the Primary Shape Inducing Conditions/Chemicals.

SYNTHETIC APPROACH	PRODUCT TYPE (PLASMONIC ABSORPTION)	LENGTH, WIDTH, ASPECT RATIO	PRIMARY SHAPE DIRECTING AGENT/CONDITION#	REFERENCE (YEAR)
Colloidal seeding growth of Au sphere	nanorod (600–1300 nm)	80–200 nm, 15–85 nm, 1.5–7	CTAB, sodium oleate, CTAC, salicylate, Ag^+	19–21 (2012, 2013)
Colloidal seeding growth of Au sphere	long nanorod (>1000 nm)	80–200 nm, 12–24 nm, 7–10	CTAB, Ag^+, hydroquinone	22 (2013)
Colloidal seeding growth of Au sphere	nanowire (>1300 nm)	1–4 micron, 30–50 nm, 30–80	CTAB, HNO_3	23 (2013)
Colloidal seedless approach	nanowire (>1400 nm)	100–150 nm, 7–13 nm, 7–20	CTAB, Ag^+, HCl	24 (2014)
Colloidal seeding growth of Au nanorod	sphere, spheroid, other shapes (600–1000 nm)	50–100 nm, 20–50 nm, 2–3	Ag underpotential deposition	25 (2016)
Colloidal seedless approach	nanorod (600–1000 nm)	40–200 nm, 20–50 nm, 2–5	CTAB, 5-bromosalicylic acid, Ag^+	26 (2017)
Colloidal seeding growth of Au nanorod	nanorod (600–1270 nm)	30–150 nm, 8–60 nm, 2–10	CTAB-decanol, Ag^+	28 (2019)
Colloidal seeding growth of Au nanorod	nanowire	5–16 µm, 20–50 nm, 200–350	CTAB, HCl	30 (2020)

(Continued)

TABLE 2.1 (CONTINUED)   History of Au Nanorod Synthetic Approaches That Are Gradually Developed and the Primary Shape Inducing Conditions/Chemicals.

SYNTHETIC APPROACH	PRODUCT TYPE (PLASMONIC ABSORPTION)	LENGTH, WIDTH, ASPECT RATIO	PRIMARY SHAPE DIRECTING AGENT/CONDITION[#]	REFERENCE (YEAR)
Multistep colloidal seeding growth	homogeneous nanorod (600–900 nm)	60–100 nm, 15–50 nm, 2–4	Ag+, CTAB	33 (2021)
Colloidal seeding growth of Au sphere	nanorod (600–1400 nm)	70–150 nm, 20–30 nm, 2–7	Varying CTAC to CTAB ratio	34 (2021)
Colloidal seeding growth of Au sphere	nanorod (600–1000 nm)	40–200 nm, 6.5 nm to 175 nm, 2–5	CTAB, Ag+, heptanol	32 (2021)

[#] hexadecyltrimethylammonium bromide (CTAB), hexadecyltrimethylammonium chloride (CTAC), benzyldimethylhexadecylammoniumchloride (BDAC), tetraoctylammonium bromide (TCAB)

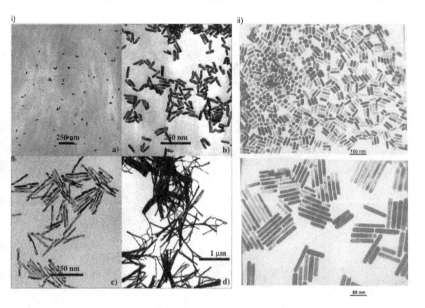

FIGURE 2.1 TEM image of Au nanorods prepared using an electrochemical approach in a hard porous template (ia-id) or in a solution-electrode interface (ii). i) Reprinted with permission from van der Zande, B. M. I. et al. 2000. Colloidal dispersions of gold rods: Synthesis and optical properties. *Langmuir*, 16, 451–468. Copyright 2000 American Chemical Society. ii) Reprinted with permission from Yu, S.-S. C et al. 1997. Gold nanorods: Electrochemical synthesis and optical properties. *The Journal of Physical Chemistry B*, 101, 6661–6664. Copyright 1997 American Chemical Society.

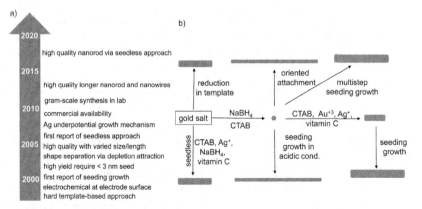

FIGURE 2.2 a) Timeline for the development of Au nanorod synthesis that leads to the commercial product development and wider availability. b) Synthetic approaches for Au nanorod that primarily focus on seeding growth in addition to seedless and other approaches.

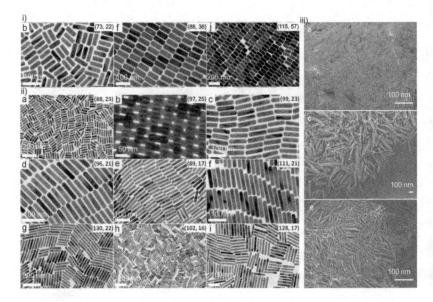

FIGURE 2.3 TEM image of Au nanorods of different lengths (ib,f,j, iia-i) and nanowires (iiia,c,e) that are prepared via the seeding growth approach. i) Reprinted with permission from Ye, X. et al. 2013. Seeded growth of monodisperse gold nanorods using bromide-free surfactant mixtures. *Nano Letters*, 13, 2163–2171. Copyright 2013 American Chemical Society. ii) Reprinted with permission from Ye, X. et al. 2013. Using binary surfactant mixtures to simultaneously improve the dimensional tuneability and monodispersity in the seeded growth of gold nanorods. *Nano Letters*, 13, 765–771. Copyright 2013 American Chemical Society. iii) Reprinted with permission from Wang, Y.-N. et al. 2013. Seed-mediated growth of ultralong gold nanorods and nanowires with a wide range of length tuneability. *Langmuir*, 29, 10491–10497. Copyright 2013 American Chemical Society.

improved during 2003 to 2008 and continues to be used presently for commercial availability of gold nanorods.[13–16,19–23,27–34]

Chemical approaches often produce colloidal forms of gold nanorod. This form of nanorod is most popular as it has a brilliant optical property that can be used for various applications. Collidal forms of gold nanorods can be processed for making film, assembled into 2D/3D structures, and used in surface chemistry exercises.[1–7] Moreover, colloidal gold nanorods in water are used in biomedical applications, such as cell labeling/targeting/therapy, because biological compartments can access them. This chapter will discuss the synthetic development of colloidal gold nanorods and the associated chemistries involved.

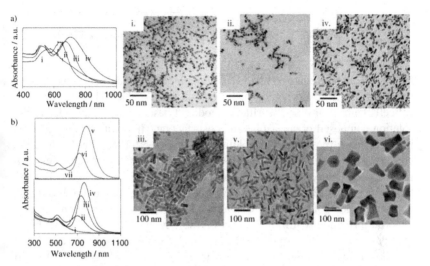

FIGURE 2.4 Optical property (ai-iv, bi-vii) and TEM image (ai,ii,vi and biii,v,vi) of gold nanorods that are prepared via the seedless approach. Reprinted with permission from Jana, N. R. 2005. Gram-scale synthesis of soluble, near-monodisperse gold nanorods and other anisotropic nanoparticles. *Small*, 1, 875–882. Copyright 2005 Wiley.

2.2 TEMPLATE-BASED SYNTHESIS

These "hard-template" syntheses offer the first opportunity to study the optical properties of gold nanorods. The template-based approach uses electrochemical depositions of gold inside a solid porous template.[9] Typically, porous alumina or polycarbonate membranes are used as templates. First, a small amount of Ag/Cu is sputtered onto the template to make the template conductive and suitable for electrodeposition. Next, Au is electrochemically deposited inside membrane pores. Finally, the template is removed and the nanorods are dispersed in media.

The diameter of nanorods is controlled by varying the membrane pore diameter and the length of nanorods is controlled by the extent of Au deposition inside membrane pores. A significant limitation of this approach is the nanorod yield, since templates can only be used once and making milligram scale nanorods is challenging. In addition, nanorods produced using this method have non-uniform surfaces that are linked to the nature of the templates and compromise the optical properties of nanorods. Figure 2.1 shows the typical image of gold nanorods prepared using this approach.

2.3 ELECTROCHEMICAL SYNTHESIS

This approach was conceptualized from the electrochemical synthesis of metal nanoparticles in reverse micelle. It consists of two electrode electrochemical cells with a Au-based sacrificial anode, Pt-based cathode, and electrolyte solution composed of surfactant and cosurfactant.[8] In addition, a Ag plate is inserted inside the electrolyte solution. During electrolysis, the Au anode is converted to $AuBr_4^-$ in a solution that migrates to the Pt cathode, reduces back to Au^0, and produces nanorods either at the cathode surface or in a bulk solution. A part of $AuBr_4^-$ oxidizes Ag^0 and releases Ag^+ in a solution that is also responsible for gold nanorod formation. However, complete mechanism of nanorod formation is still unsolved. Interestingly this method produces high-quality cylindrical nanorods and the length of the nanorods can be controlled by adjusting experimental conditions (Figure 2.1ii). Typical length, width, and aspect ratios of synthesized nanorods are 20–150 nm, 10–15 nm, and 2–10, respectively (see Table 2.1). Although this approach is the first one that produced high-quality gold nanorods, the method is relatively complex and synthesis at a scale of a few milligrams requires enormous effort.

2.4 SEED-MEDIATED SYNTHESIS

This is a simple wet chemical approach that offers convenient and reliable synthesis of colloidal gold nanorods with controllable length to width ratios. This approach was conceptualized in 2001 from four different reports/aspects at that time.[35–38] First, CdSe semiconductor nanorods with controlled length were synthesized using an isotropic solution-based approach by maintaining a high monomer concentration (this induces better stability of anisotropic nanoparticle embryos) and overdriven (faster than usual cases) growth kinetics of nucleation centers.[35] Although this method inspired gold nanorod synthesis, the question arises of whether the successful synthesis of crystallographically asymmetric bulk CdSe systems can be adapted for highly symmetric cubic gold systems. Second, seeding growth has been extensively explored for synthesis of highly monodispersed spherical gold nanoparticles in the size range 20–100 nm. It has been noticed that faster growth occurs under seeding condition and interestingly a small population of gold nanorods are formed in certain experimental conditions.[36] These experimental observations suggest that gold nanorods may be obtained in higher yields by manipulating seeding growth conditions. Third, nucleation-growth kinetics in solution-chemical gold nanoparticle synthesis is coupled with autocatalytic reduction of gold salt that induces secondary nucleation during the seeding growth.[37] This restricts the wider applicability of the seeding growth approach for shape control, including with the use of high monomer concentration. The identification of a seeding growth condition using hexadecyltrimethylammonium bromide (CTAB) as a surfactant stabilizer and vitamin C as a weak reducing agent that minimize such secondary nucleation offers potential to explore wider growth conditions for making anisotropic gold nanoscructures[38] Fourth, wider exploration of seeding growth of gold nanoparticles in CTAB/vitamin C has identified certain selected conditions that modify growth mechanisms and enhance the yield of gold nanorods.[37,38] These conditions include <5 nm seed size, higher CTAB concentration (>0.1 M that is above 2nd cmc (critical micelle concentration) that forms cylindrical micelle), additives that are known to induce cylindrical micelle formation, and presence of silver salt.

The first effective gold nanorod synthesis via seeding growth was reported in two papers in 2001.[10,11] In one work, cylindrical gold nanorods with a 5 to 18 aspect ratio are formed.[10] In this method, nanorods are actually formed as a side product (<5 % yield) along with nanospheres as major products. The nanorods are shape separated. This method does not use silver salt. Other methods report spheroidal and elongated rod-like products of varied aspect ratios from 1.5 to 5 and with a 50–60 % yield. This method used silver salt and CTAB to enhance the yield of nanorods. These initial methods have been modified several times to improve nanorod yield and quality. Figure 2.2a and Table 2.1 summarize the timelines for the development and gradual improvement of Au nanorod synthesis that finally led to their commercial availability and wider use. Important developments include fine tuning the silver salt concentration and <3 nm seed for enhanced nanorod yield,[13–17] and the use of small molecule additives to prepare high-quality small nanorods,[13,19–21] multistep seeding to increase the yield of longer nanorods,[33] and acidic conditions to enhance nanowire yield.[23]

A typical seeding growth method (that is currently used) is shown in Figure 2.2b. The process involves the growth of <3 nm seed in an aqueous growth solution that contains CTAB, gold salt ($AuCl_4^-$), silver salt ($AgNO_3$), and vitamin C. A rod-like shape is induced from the combined effect of CTAB and Ag^+. Vitamin C plays two roles. First, it instantaneously reduces all Au^{+3} to Au^+. Second, it reduces Au^+ to Au^0 in the presence of an Au seed, which means only at the surface of an Au seed. Once the seed is added to the growth solution, the <3 nm Au seeds anisotropically grow into nanorods structures. The length of nanorods are primarily determined by seed to gold salt ratio. In addition, the concentration of silver salt and CTAB, seed size, other small molecule additives, multistep seeding, and solution pH are used to control the length to width ratio, and the width of nanorods. The seeding growth method is now significantly advanced and colloidal gold nanorods of different length and aspect ratios are commercially available. Examples of some high-quality nanorods are shown in Figure 2.3.

2.5 SEEDLESS SYNTHESIS

With the advancement of seeding growth approaches, it was realized that smaller seeds are better for shape control. In particular, seeds that are 1.5 nm or smaller offer better quality nanorods.[13] As the seed size is increased above 5 nm the symmetry breaking appears difficult and the yield of elongated nanorod structures becomes lower.[17] However, it is very challenging to prepare seeds at the ideal size of <3 nm and adjust their concentration in order to make nanorods of a specific length. This is because seeds should be stable enough without any aggregation or enlargement, until nanorod growth is initiated. In addition, the seed's surface should be available for additional deposition of gold for nanorod growth. This is a difficult issue as stabilizer is required for seed via capping around the seed but those stabilizers can inhibit nanorod growth. In practice, CTAB is often used as a capping agent for <3 nm seed because it allows seeds to grow into nanorod structures.[13–15] However, CTAB-capped gold seeds grow in size with time and also change the crystal structure before they are used for nanorod growth. This leads to poor reproducibility of nanorod length and aspect ratios. Thus, it is desirable to replace the two-step approach to seeding growth but without compromising the quality of gold nanorods.

As a result, a seedless synthesis of gold nanorods was designed and first reported in 2005.[17] In this approach, the two-step approach to seeding growth is replaced by one-step and one-pot approach (Figure 2.2b). Typically, an aqueous solution is prepared that contains CTAB, gold salt ($AuCl_3^-$), and silver salt ($AgNO_3$) at required concentrations. Next, a mixture of borohydride and vitamin C is added under a stirring condition. Borohydride is a strong reducing agent that primarily induces seed nucleation and vitamin C (a weak reducing agent) induces the growth of seeds into nanorod structures. The length of nanorods is primarily determined by the ratio of borohydride to vitamin C. Other factors that control nanorod length and aspect ratios include concentration of silver salt and CTAB. Figure 2.4 shows nanorods of controlled length to width ratios that can be prepared by this approach. One important advantage of this approach is that nanorods with smaller widths (<10 nm) can be prepared. However, the length to width ratios of nanorods can only be controlled in a short range (typically up to 5 of length to width ratio). This method has been further improved in recent years to make better quality nanorods and to increase the aspect ratio of prepared nanorods.[24,26]

2.6 LONG NANOROD AND NANOWIRE SYNTHESIS

Although high-quality nanorods of shorter length can be prepared, the synthesis of longer nanorods and nanowires require additional attention. In particular, the synthesis of nanorods with aspect ratios >20 and plasmonic absorption maximum of >1300 nm and micron size nanowires are difficult to prepare. There are two reasons for this. First, long nanorod and nanowire syntheses require longer growth periods, which increases the possibility of secondary nucleation. Second, the nanorod growth conditions that are used to restrict width dimension (via use of silver salt and CTAB) and allow one-dimensional growth. These conditions can restrict the overall growth over an extended period with a resultant inhibition of nanowire structures.

Thus further manipulation of growth conditions and alternative growth approaches have been explored. The most successful approaches that produce longer nanorods and nanowires include oriented attachment of spherical seeds[18,23] or small nanorods[30] using modified growth conditions. Modified growth conditions include organic medium or acidic growth conditions (Figure 2.2b). For example, growth conditions are adjusted for oriented attachment of spherical gold seeds to form micron-length gold nanowires that have a 2 nm diameter.[18] Similarly, the seeding growth condition is adjusted in acidic media for oriented attachment of Au spheres or Au nanorods.[23,30] These approaches allow the formation of several micron-length gold nanorod that are 20–50 nm in diameter. Alternatively, the seedless approach has been adapted in acidic growth conditions to produce gold wires 100–150 nm in length with diameters 7–13 nm.[24]

2.7 FORMATION MECHANISM

Although there is significant progress in the synthesis of high-quality gold nanorods, the exact mechanism of formation is still under debate. This is particularly because of complex growth conditions and the involvement of multiple factors. For example, CTAB can act either as a capping agent or as a micellar template, or both. Silver salt can form colloidal AgBr nanoparticles capped with CTAB or soluble $CTA^+AgBr_2^-$-based ion associate, depending on the molar ratio of CTAB to silver salt. The $AuCl_4^-$ can be partly converted to $AuBr_4^-$ or form $CTA^+AuBr_4^-$-based ion associate. Moreover, the presence of vitamin C can reduce $AuCl_4^-$ to $AuCl_2^-$ and form similar type ion associates. All of these factors can influence and complicate the anisotropic growth mechanism.

Considering all these aspects, four different mechanisms have been proposed and at least one of them is primarily involved in a selected growth condition.[2-6,39-45] Figure 2.5 summarizes these four mechanisms. The first one is the Ag underpotential deposition with the assistance of CTAB.[6,40] This mechanism induces symmetry breaking using an underpotential deposition of Ag monolayer, preferentially at {110} face of growing seed. This preferential deposition is possibly assisted via the capping of growing particles by $CTA^+AgBr_2^-$. This process protects the {110} side face and allows growth along the un-passivated {111} face. This mechanism can explain the silver-assisted growth of gold nanorods with aspect ratios 1–10.

The second one is the zipping mechanism which combines CTAB with Ag^+ and other additives.[3] According to this mechanism, CTAB preferentially caps the {110} side face and offers one-dimensional growth with the formed nanorods having {110} side face and {111} end face. This mechanism also explains the silver-assisted growth of gold nanorods with aspect ratios 1–10. The third one is the zipping of growing particles using a surfactant bilayer or lamer phase.[45] This mechanism may or may not require Ag^+. This mechanism can explain the formation of longer nanorods and nanowires via oriented attachment or multistep seeding approach. Fourth is the template mechanism in a rod-like surfactant micelle.[17] According to this mechanism, seed nanoparticles grow at the surface of rod-like micelle that induce symmetry breaking. At the end stage, nanorod growth occurs in a manner similar to surfactant zipping or the underpotential Ag deposition approach. This mechanism is most convenient for the seedless approach and creates nanorods with a <5 aspect ratio. Further studies are ongoing to converge approaches and develop a generalized mechanism that will predict nanorod dimension and quality and provide suitable growth conditions.

2.8 PURIFICATION OF GOLD NANOROD

All synthesized nanorods are mixed with an excess amount of CTAB and some other reagents. In certain synthetic methods, they are also often mixed

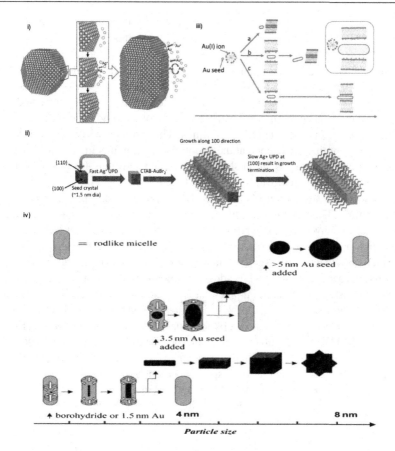

FIGURE 2.5 Anisotropic growth mechanisms that are proposed in the formation of gold nanorods/nanowires. They include Ag underpotential deposition by the assistance of CTAB (i), surfactant zipping mechanism with the assistance of AgBr and aromatic additives (ii), zipping of growing particles inside surfactant bilayer or lamer phase (iii), and rod-like surfactant micelle template (iv). i) Reprinted with permission from Walsh, M. J. et al. 2017. A mechanism for symmetry breaking and shape control in single crystal gold nanorods. *Accounts of Chemical Research*, 50, 2925–2935. Copyright 2017 American Chemical Society. ii) Reprinted with permission from Murphy, C. J. et al. 2011. Gold nanorod crystal growth: From seed-mediated synthesis to nanoscale sculpting. *Current Opinion in Colloid & Interface Science*, 16, 128–134. Copyright 2011 Elsevier. iii) Reprinted with permission from Inaba, T. et al. 2019. Effect of the crystallization process of surfactant bilayer lamellar structures on the elongation of high-aspect-ratio gold nanorods. *The Journal of Physical Chemistry B*, 123, 4776–4783. Copyright 2019 American Chemical Society. iv) Reprinted with permission from Jana, N. R. 2005. Gram-scale synthesis of soluble, near-monodisperse gold nanorods and other anisotropic nanoparticles. *Small*, 1, 875–882. Copyright 2005 Wiley.

with spherical side products. Thus it is critical to purify nanorods from excessive CTAB and separate them from spherical gold nanoparticles prior to their use. The separation of nanorods from CTAB and other reagents is performed primarily via high-speed centrifuge. Typically, colloidal nanorods are centrifuged to precipitate CTAB-capped nanorods. Next, colorless supernatant with excess CTAB is discarded and precipitated nanorods are redispersed in fresh water. This centrifuge-based precipitation of nanorods and redispersion in fresh water can be repeated 3–4 times. This process will remove excess CTAB, except that are capped at the nanorod's surface. It is important to mention that excessive purification should be avoided as it may lead to insoluble nanorods due to the removal of capped CTAB.

The shape separation of nanorods from spherical side products is very challenging. This is particularly due to the similar size of nanorods and spheres and conventional centrifuge or membrane filtration is inefficient. Thus scientists have explored alternative approaches. One successful approach is the surfactant-assisted self-assembly via depletion attraction-based approach.[46,47] In this approach, excess CTAB is added to colloidal nanorods and temperature/time is adjusted for preferential precipitation of nanorods. Under this condition nanorods preferentially assemble into liquid crystal-like structures and precipitate from solution. This is possible due to the molecular crowding effect imposed by CTAB and enhances the nanorod–nanorod interaction via capped CTAB. Nanorods can be isolated and dispersed in fresh CTAB solution. This type of precipitation and redispersion can be repeated 2–3 times for the complete removal of spherical side products. Finally, excess CTAB can be removed using the approach described above. This approach is first reported 2003[46] and further advanced in later years.[47]

REFERENCES

1. Perez-Juste´, J., Pastoriza-Santos, I., Liz-Marzan´, L. M. and Mulvaney, P. 2005. Gold nanorods: Synthesis, characterization and applications. *Coordination Chemistry Reviews*, 249, 1870–1901.
2. Huang, X., Neretina, S. and El-Sayed, M. A. 2009. Gold nanorods: From synthesis and properties to biological and biomedical applications. *Advanced Materials*, 21, 4880–4910.
3. Murphy, C. J., Thompson, L. B., Chernak, D. J., Yang, J. A., Sivapalan, S. T., Boulos, S. P., Huang, J.; Alkilany, A. M. and Sisco, P. N. 2011. Gold nanorod crystal growth: From seed-mediated synthesis to nanoscale sculpting. *Current Opinion in Colloid & Interface Science*, 16, 128–134.

4. Vigderman, L., Khanal, B. P. and Zubarev, E. R. 2012. Functional gold nanorods: Synthesis, self-assembly, and sensing applications. *Advanced Materials*, 24, 4811–4841.
5. Lohse, S. E. and Murphy, C. J. 2013. The quest for shape control: A history of gold nanorod synthesis. *Chemistry of Materials*, 25, 1250–1261.
6. Walsh, M. J., Tong, W., Katz-Boon, H., Mulvaney, P., Etheridge, J. and Funston, A. M. 2017. A mechanism for symmetry breaking and shape control in single crystal gold nanorods. *Accounts of Chemical Research*, 50, 2925–2935.
7. Zheng, J., Cheng, X., Zhang, H., Bai, X., Ai, R., Shao, L. and Wang, J. 2021. Gold nanorods: The most versatile plasmonic nanoparticles. *Chemical Reviews*, 121, 13342–13453.
8. Yu, S.-S. C., Lee, C. L. and Wang, C. R. C. 1997. Gold nanorods: Electrochemical synthesis and optical properties. *The Journal of Physical Chemistry B*, 101, 6661–6664.
9. van der Zande, B. M. I., Bo¨hmer, M. R., Fokkink, L. G. and Scho¨nenberger, C. 2000. Colloidal dispersions of gold rods: Synthesis and optical properties. *Langmuir*, 16, 451–458.
10. Jana, N. R., Gearheart, L. and Murphy, C. J. 2001. Wet chemical synthesis of high aspect ratio cylindrical gold nanorods. *The Journal of Physical Chemistry B*, 105, 4065–4067.
11. Jana, N. R., Gearheart, L. and Murphy, C. J. 2001. Seed-mediated growth approach for shape-controlled synthesis of spheroidal and rod-like gold nanoparticles using a surfactant template. *Advanced Materials*, 13, 1389–1393.
12. Kim, F., Song, J. H. and Yang, P. 2002. Photochemical synthesis of gold nanorods. *Journal of the American Chemical Society*, 124, 14316–14317.
13. Nikoobakht, B. and El-Sayed, M. A. 2003. Preparation and growth mechanism of gold Nanorods (NRs) using seed-mediated growth method. *Chemistry of Materials*, 15, 1957–1962.
14. Sau, T. K. and Murphy, C. J. 2004. Room temperature, high-yield synthesis of multiple shapes of gold nanoparticles in aqueous solution. *Journal of the American Chemical Society*, 126, 8648–8649.
15. Sau, T. K. and Murphy, C. J. 2004. Seeded high yield synthesis of short Au nanorods in aqueous solution. *Langmuir*, 20, 6414–6420.
16. Gole, A. and Murphy, C. J. 2004. Seed-mediated synthesis of gold nanorods: Role of the size and nature of the seed. *Chemistry of Materials*, 16, 3633–3640.
17. Jana, N. R. 2005. Gram-scale synthesis of soluble, near-monodisperse gold nanorods and other anisotropic nanoparticles. *Small*, 1, 875–882.
18. Halder, A. and Ravishankar, N. 2007. Ultrafine single-crystalline gold nanowire arrays by oriented attachment. *Advanced Materials*, 19, 1854–1858.
19. Ye, X., Jin, L., Caglayan, H., Chen, J., Xing, G., Zheng, C., Doan-Nguyen, V., Kang, Y., Engheta, N., Kagan, C. R. and Murray, C. B. 2012. Improved size-tunable synthesis of monodisperse gold nanorods through the use of aromatic additives. *ACS Nano*, 6, 2804–2817.
20. Ye, X., Zheng, C., Chen, J., Gao, Y. and Christopher B. Murray, C. B. 2013. Using binary surfactant mixtures to simultaneously improve the dimensional tunability and monodispersity in the seeded growth of gold nanorods. *Nano Letters*, 13, 765–771.

21. Ye, X., Gao, Y., Chen, J., Reifsnyder, D. C., Zheng, C. and Murray, C. B. 2013. Seeded growth of monodisperse gold nanorods using bromide-free surfactant mixtures. *Nano Letters*, 13, 2163–2171.
22. Vigderman L. and Zubarev, E. R. 2013. High-yield synthesis of gold nanorods with longitudinal SPR peak greater than 1200 nm using hydroquinone as a reducing agent. *Chemistry of Materials*, 25, 1450–1457.
23. Wang, Y.-N., Wei, W.-T., Yang, C.-W. and Huang, M. H. 2013. Seed-mediated growth of ultralong gold nanorods and nanowires with a wide range of length tunability. *Langmuir*, 29, 10491–10497.
24. Xu, X., Zhao, Y., Xue, X., Huo, S., Chen, F., Zou, G. and Liang, X.-J. 2014. Seedless synthesis of high aspect ratio gold nanorods with high yield. *Journal of Materials Chemistry A*, 2, 3528–3535.
25. Zhang, Q., Jing, H., Li, G. G., Lin, Y., Blom, D. A. and Wang, H. 2016. Intertwining roles of silver ions, surfactants, and reducing agents in gold nanorod overgrowth: Pathway switch between silver underpotential deposition and gold–silver codeposition. *Chemistry of Materials*, 28, 2728–2741.
26. Liu, K., Bu, Y., Zheng, Y., Jiang, X., Yu, A. and Wang, H. 2017. Seedless synthesis of monodispersed gold nanorods with remarkably high yield: Synergistic effect of template modification and growth kinetics regulation. *Chemistry – A European Journal*, 23, 3291–3299.
27. Chang, H.-H. and Murphy, C. J. 2018. Mini gold nanorods with tunable plasmonic peaks beyond 1000 nm. *Chemistry of Materials*, 30, 1427–1435.
28. Gonzalez-Rubio, G., Kumar, V., Llombart, P., Díaz-Núñez, P., Bladt, E., Altantzis, T., Bals, S., Peña-Rodríguez, O., Noya, E. G., MacDowell, L. G., Guerrero-Martínez, A. and Liz-Marzan, L. M. 2019. Disconnecting symmetry breaking from seeded growth for the reproducible synthesis of high quality gold nanorods. *ACS Nano*, 13, 4424–4435.
29. Khanal, B. P. and Zubarev, E. R. 2019. Gram-scale synthesis of isolated monodisperse gold nanorods. *Chemistry – A European Journal*, 25, 1595–1600.
30. Khanal, B. P. and Zubarev, E. R. 2020. Gold nanowires from nanorods. *Langmuir*, 36, 15030–15038.
31. Wei, M.-Z., Deng, T.-S., Zhang, Q., Cheng, Z. and Li, S. 2021. Seed-mediated synthesis of gold nanorods at low concentrations of CTAB. *ACS Omega*, 6, 9188–9195.
32. He, H., Wu, C., Bi, C., Song, Y., Wang, D. and Xia, H. 2021. Synthesis of uniform gold nanorods with large width to realize ultralow SERS detection. *Chemistry – A European Journal*, 27, 7549–7560.
33. Zhang, X., Tran, N., Egan, T., Sharma, B. and Chen, G. 2021. Synthesis of homogeneous gold nanorods through the optimized multi-step seed-mediated growth method. *The Journal Physical Chemistry C*, 125, 13350–13360.
34. Sánchez-Iglesias, A., Jenkinson, K., Bals, S. and Liz-Marzán, L. M. 2021. Kinetic regulation of the synthesis of pentatwinned gold nanorods below room temperature. *The Journal Physical Chemistry C*, 125, 23937–23944.
35. Peng, X., Manna, L., Yang, W., Wickham, J., Scher, E., Kadavanich, A. and Alivisatos, A. P. 2000. Shape control of CdSe nanocrystals. *Nature*, 404, 59–61.
36. Brown, K. R., Walter, D. G. and Natan, M. J. 2000. Seeding of colloidal Au nanoparticle solutions. 2. Improved control of particle size and shape. *Chemistry of Materials*, 12, 306–313.

37. Jana, N. R., Gearheart, L. and Murphy, C. J. 2001. Evidence for seed-mediated nucleation in the chemical reduction of gold salts to gold nanoparticles, *Chemistry of Materials*, 13, 2313–2322.
38. Jana, N. R., Gearheart, L. and Murphy, C. J. 2001. Seeding growth for size control of 5–40 nm diameter gold nanoparticles. *Langmuir*, 17, 6782–6786.
39. Johnson, C. J., Dujardin, E., Davis, S. A., Murphy, C. J. and Mann. S. 2002. Growth and form of gold nanorods prepared by seed-mediated, surfactant-directed synthesis. *Journal of Materials Chemistry*, 12, 1765–1770.
40. Liu, M. and Guyot-Sionnest, P. 2005. Mechanism of silver (I)-assisted growth of gold nanorods and bipyramids. *The Journal of Physical Chemistry B*, 109, 22192–2220.
41. Meena, S. K. and Sulpizi, M. 2016. From gold nanoseeds to nanorods: The microscopic origin of the anisotropic growth. *Angewandte Chemie International Edition*, 55, 11960–11964.
42. Park, K., Hsiao, M.-S., Koerner, H., Jawaid, A., Che, J. and Vaia, R. A. 2016. Optimizing seed aging for single crystal gold nanorod growth: The critical role of gold nanocluster crystal structure. *The Journal of Physical Chemistry C*, 120, 28235–28245.
43. Tong, W., Walsh, M. J., Mulvaney, P., Etheridge, J. and Funston, A. M. 2017. Control of symmetry breaking size and aspect ratio in gold nanorods: Underlying role of silver nitrate. *The Journal of Physical Chemistry C*, 121, 3549–3559.
44. da Silva, J. A. and Meneghetti, M. R. 2018. New aspects of the gold nanorod formation mechanism via seed-mediated methods revealed by molecular dynamics simulations. *Langmuir*, 34, 366–375.
45. Inaba, T., Takenaka, Y., Kawabata, Y. and Kato, T. 2019. Effect of the crystallization process of surfactant bilayer lamellar structures on the elongation of high-aspect-ratio gold nanorods. *The Journal of Physical Chemistry B*, 123, 4776–4783.
46. Jana, N. R. 2003. Nanorod shape separation using surfactant assisted self-assembly. *Chemicals Communications*, 1950–1951.
47. Park, K., Koerner, H. and Vaia, R. A. 2010. Depletion-induced shape and size selection of gold nanoparticles. *Nano Letters*, 10, 1433–1439.

Gold Nanorod-Based Different Nanocomposites

3

3.1 INTRODUCTION

Gold nanorod-based different nanocomposites are primarily prepared to support the more effective use of the plasmonic properties of gold nanorods in diverse applications, including optical probes, bioimaging, photoredox catalysis, solar energy conversion, etc.[1–3] In particular, multimodal approaches combined with different nanomaterials can offer advantages in synergistic therapeutic properties. For example, the coupling of conventional chemotherapy with photodynamic therapy offered by gold nanorods provides more efficient therapy with lower side effects.[2] Similarly, coupling with fluorescent nanoparticles or dyes can offer dark field and fluorescence-based dual imaging, coupling with magnetic nanoparticles can offer dark field and magnetic resonance imaging (MRI)-based dual imaging, coupling with catalytic metal/metal oxide can offer multimodal/enhanced catalysis, and coupling with conducting polymer can enhance photocatalytic and solar energy conversion performance.[1,3]

Table 3.1 summarizes the gold nanorod-based different nanocomposites that are prepared using a wide variety of heterostructures. It summarizes their composition, properties, and application potentials. Diverse strategies have been used to make those structures, including chemical conjugation, electrostatic interaction, coating, hydrophobic interaction, and encapsulation inside a porous shell. The size of nanocomposites varies widely from 50 to 500 nm, depending on the synthetic strategy used and components present. The application potential can also vary. They can be used in responsive drug delivery, photothermal therapy, biosensing, bioimaging, cell therapy, tumor therapy, catalysis, and solar energy conversion. In this chapter, we discuss these nanocomposites based on structural components other than gold nanorods and include composites with other nanoparticles, inorganic material-coated gold

DOI: 10.1201/9781003245339-3

TABLE 3.1   Summary of Gold Nanorod-Based Composites, Their properties, and Application Potential.

NANOCOMPOSITE	PROPERTY	SIZE	APPLICATION POTENTIAL	REFERENCE
Gold nanorod-quantum dot	enhanced photoluminescence of quantum dot	50–100 nm	bioimaging	4
Gold nanorod-Fe_3O_4-hydrogel	multi-responsive property	–	responsive delivery and therapy	5
TiO_2-gold nanorod-indocyanine green	plasmon-enhanced ROS generation, photothermal drug delivery	100–200 nm	photothermal catalysis/ therapy,	6, 9, 16
Gold nanorod-upconversion nanoparticle- photosensitizer dye	light-induced ROS generation	50–100 nm	photothermal and photodynamic therapy	8
Gold nanorod-silicon nanoparticle	plasmon enhanced luminescence of silicon nanoparticle	100 nm	bioimaging	10
Ceria-coated gold nanorod	NIR light-induced Fenton reaction and current generation	50–100 nm	light-induced therapy	11
Gold nanorod-copper sulfide	enhanced photothermal efficiency	100–150 nm	cell therapy	12
Platinum-coated gold nanorod	efficient ROS scavengers	25–50 nm	prevent oxidative damage during phototherapy	13
Gold-palladium nanorod	tunable catalysis	100–200 nm	catalysis	14
Silver-coated gold nanorod	enhanced ROS generation	50–100 nm	cell therapy via oxidative stress	15
Gadolinium oxysulfide-coated gold nanorod	dual-modal magnetic resonance and photoacoustic imaging	50–100 nm	cancer imaging and therapy	17

(Continued)

TABLE 3.1 (CONTINUED) Summary of Gold Nanorod-Based Composites, Their properties, and Application Potential.

NANOCOMPOSITE	PROPERTY	SIZE	APPLICATION POTENTIAL	REFERENCE
Gold nanorod-layered double hydroxide	efficient photothermal effect	100–500 nm	antibacterial and tumor therapy	18
Polydopamine-coated gold nanorod	drug delivery carrier, photoacoustic imaging probe	100 nm	imaging guided chemo-photothermal therapy	19
Cu(II)-doped polydopamine-coated gold nanorod	enhanced photothermal performance	50–100 nm	phototherapy and chemotherapy	20
Gold nanorod-metal organic framework	combinational phototherapy	100–500 nm	tumor therapy, biosensing	21, 22
Gold nanorod-conjugated polymer	enhanced colloidal stability and electron transport	100–150 nm	ink for printed electronic	23
Gold nanorod-conducting polymer	enhanced photothermal property	100–200 nm	photothermal therapy	24
Gold nanorod-dye	coherent coupling between exciton and surface plasmon	25–50 nm	quantum optical studies	26
Graphene-hemin-gold nanorod	higher catalytic ROS generation	>500 nm	biosensing	29
Gold nanorod-porous silicon	drug delivery carrier	200–300 nm	photothermal and combination therapy	30
Gold-nanorod-platinum	photoresponsivity	100–200 nm	infrared photodetector	32

nanorods, organic polymer-coated gold nanorods, small molecule-based composites, and other multifunctional nanocomposites.

3.2 COMPOSITES WITH OTHER NANOPARTICLES

Gold nanorod-based composites with different nanoparticles have been prepared to enhance the application potential of gold nanorods.[4–10] The conjugated nanoparticles include quantum dots, upconversion nanoparticles, fluorescent silicon nanoparticles, magnetic nanoparticles, and fluorescent gold clusters. In most cases, these nanoparticles are prepared separately and then conjugated with previously prepared gold nanorods via physical or chemical approaches. Alternatively, preformed gold nanorod surface is used to nucleate other nanoparticles or other nanoparticles are incorporated into polymer shells around nanorods.

One interesting direction is the enhancement of the fluorescence property of conventional fluorescent nanoparticles via conjugation with gold nanorods. For example, the localized plasmons of gold nanorods are used to enhance the luminescence properties of quantum dots and silicon nanoparticles.[4, 10] Quantum dots are conjugated with gold nanorods to modulate their emission properties via interaction with the longitudinal localized surface plasmon resonance of the nanorods. Anionic quantum dots are electrostatically assembled around cationic gold nanorods with varying spacers between them. It has been observed that the emission of quantum dots is reduced if spacing between them is <5 nm. This is due to the efficient metal-induced energy transfer. In contrast, if a spacer is used to make a distance >5 nm, the quantum dot emission is enhanced 5 times under the plasmonic excitation of nanorods via a plasmon-exciton interaction.[4]

Similarly, core-shell-like gold nanorod-silicon nanocrystals with silica shells as tunable spacers have been synthesized.[10] It is observed that a separation distance of 5 nm increases the luminescence intensity of silicon nanoparticles by 7 times.[10] Longitudinal plasmonic bands of nanorods also play a prominent role in luminescence enhancement. This has been explained by plasmon–luminescence coupling, where longitudinal plasmonic bands of nanorods were used to determine the photoluminescence properties of silicon nanocrystals using the Purcell-enhanced radiative rate effect. In other work, gold nanorods were surface decorated with fluorescent gold clusters to prepare plasmonic-fluorescent nanoprobes.[7] In other work, lanthanide-doped upconversion nanoparticles were electrostatically bound with mesoporous silica-coated gold nanorods.[8] Nanorods were selected in such a way that the fluorescence of upconversion nanoparticles could stimulate nanorod plasmon to generate heat through energy transfer. Loading of photosensitizer inside mesoporous silica can generate reactive oxygen species (ROS), which is further enhanced through the surface plasma resonance effect of nanorods.[8]

The positioning of photoactive species at either end of gold nanorods is an important part of utilizing the electric field enhancement effect at those ends. As a result, gold nanorods with porous TiO_2 caps at either end have been synthesized.[6,9] Figure 3.1 shows an example of such nanocomposites. Photosensitizer is loaded at either end and used for plasmon-enhanced ROS generation. It has been observed that higher amounts of singlet oxygen are generated if photosensitizer is present at the tips of nanorods.[9] Similar types of

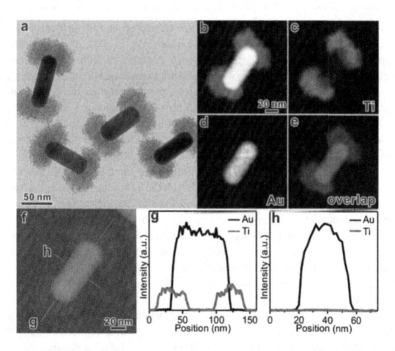

FIGURE 3.1   (a) TEM image of gold nanorod with porous TiO_2 caps at either end, (b–e) elemental mapping of individual nanocomposites, and (f–h) elemental profiles of individual components. Reprinted with permission from Wu, B. et al. 2016. Anisotropic growth of TiO_2 onto gold nanorods for plasmon-enhanced hydrogen production from water reduction. *Journal of the American Chemical Society*, 138, 1114–1117. Copyright 2016 American Chemical Society.

TiO_2-gold nanorod composites have been used to generate plasmon-enhanced hydrogen from water.[6]

Responsive hydrogels are superior for selective drug delivery applications. Gold nanorod and magnetic nanoparticle-loaded hydrogels are prepared for such applications. Here nanorod components offer NIR light-induced heat generation and magnetic nanoparticles offer magneto-responsive properties.[5] Similar types of magnetic nanocomposites are useful for multimodal imaging and therapeutic applications.[3]

3.3 INORGANIC MATERIAL-COATED GOLD NANOROD

A wide variety of inorganic materials are deposited at the surface of gold nanorods in order to produce composite nanoparticles with multifunctional properties. Compared to nanoparticle-based composites, here the inorganic materials form a continuous layer with varied thicknesses, as shown in Figure 3.2. Deposited materials can be semiconductors, metal, metal oxide, and porous silica. The final property varies depending on the deposited material and thickness (see Table 3.1 for more details). In one example, core-shell type composites were designed by depositing CeO_2 (ceria) around the gold nanorod.[11] The CeO_2 is used for the well-known redox cycle of Ce^{4+}/Ce^{3+} that offers Fenton-like reactions. The composite maintains the tunable longitudinal plasmon resonance of gold nanorods and thus NIR light can be used for plasmonic photochemistry. It has been shown that this composite can accelerate the ceria-dependent Fenton-like reaction through the plasmon-induced hot-electron injection under NIR light illumination.[11] Similar types of core-shell nanoparticles are prepared using gold nanorod cores and Cu_2S shells. These nanoparticles show enhanced photothermal efficiency and offer synergetic photothermal effects due to nanorod components and chemical dynamic therapy due to Cu_2S.[12]

In alternative approaches, other catalytic metals are deposited around gold nanorods to modulate the catalytic property. For example, platinum coated gold nanorods are prepared to scavenge ROS generated during plasmonic photothermal therapy to minimize side effects.[13] Moreover, gold nanorod has been used as a template for the deposition of palladium and to produce Au–Au alloy cores and Pd shells with multifaceted geometries. These materials offer tunable catalytic properties.[14] In another approach, silver-coated gold nanorod is designed with controlled ROS generation properties that can be used for protective autophagy in live cells.[15]

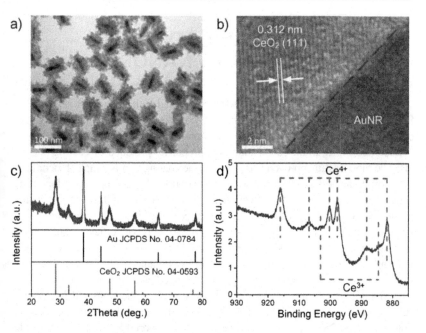

FIGURE 3.2   Example of core-shell type nanocomposite where inorganic CeO$_2$ forms a continuous layer around gold nanorod. (a) TEM image at lower resolution, (b) HRTEM image, (c) XRD pattern, and (d) Ce 3d XPS spectrum nanocomosite. Reprinted with permission from Wang, J.-H. et al. 2016. Ceria-coated gold nanorods for plasmon-enhanced near-infrared photocatalytic and photoelectrochemical performances. *The Journal Of Physical Chemistry C*, 120, 14805–14812. Copyright 2016 American Chemical Society.

Semiconducting layers are deposited around gold nanorods to create enhanced photocatalytic properties. Gold nanorods coated with a uniform and continuous layer of titanium dioxide nanoshells have been prepared. This material is used for the NIR light-induced oxidation of glucose at near room temperature. This material offers enhanced ROS generation under light exposure, as compared to pure gold nanorod, and offers glucose oxidation similar to glucose oxidase.[16] In another approach, Gd$_2$O$_2$S coated gold nanorod is prepared where nanorod components offer photoacoustic imaging and photothermal therapy applications and Gd$_2$O$_2$S components offer MRI contrast agents.[17] In another approach, core-shell material is prepared with a gold nanorod core and layered double hydroxides shell. This material offers enhanced thermal energy conversion and more efficient photothermal therapy.

3.4 ORGANIC POLYMER-COATED GOLD NANOROD

Organic polymer-coated gold nanorods are prepared with three specific goals.[18–24] First, to enhance colloidal stability and to remove toxic CTAB capping. Second, to enhance the photothermal effect offered by gold nanorods via enhanced electronic conduction. Third, to introduce additional properties provided by coating materials. The most useful approach is polydopamine coating as it is simple, biocompatible, porous, and offers easy functionalization options. For example, polydopamine coating of gold nanorod offers functionalization with cyclic RGD (arginylglycylaspartic acid) peptide and loading with anti-cancer drugs inside the porous polymer layer. This material can target tumor angiogenic vessels with efficient chemo-thermal tumor therapy.[19] In a similar approach, Cu(II)-doped polydopamine-coated gold nanorod has been synthesized that offers enhanced photothermal performance via increased biocompatibility and prolonged blood circulation time.[20]

Metal–organic frameworks (MOFs) have diverse applications and thus encapsulation of gold nanorods inside MOFs can significantly enhance the application potential of gold nanorods. To achieve this, gold nanorod is encapsulated inside zirconium-based MOFs. The MOF is able to take-up or block molecules from the pores and the core nanorod can be used for surface-enhanced Raman spectroscopic detection applications.[21] In a similar approach, a porphyrinic MOF shell is designed on the surface of gold nanorods and used for photoinduced singlet oxygen generation. This material is used for efficient tumor treatment by combining photodynamic therapy and photothermal therapy.[22]

Coating with conducting polymer can enhance the photothermal performance via enhanced electron transport from the nanorod's surface. Gold nanorod is coated with a conducting polymer that offers enhanced colloidal stability and good electron transport properties. Figure 3.3 shows an electron microscopic image of such a structure. This material is used as conductive ink for printed electronics[23] and the photothermal therapy of tumors.[24] In a similar concept, layered double hydroxide is used to coat gold nanorods for enhancing the photothermal effect and this material is used in antibacterial applications.[18]

3.5 SMALL MOLECULE-BASED COMPOSITE

A wide variety of small molecule conjugates of gold nanorods are synthesized for different applications. They include biological targeting, better detection/imaging,

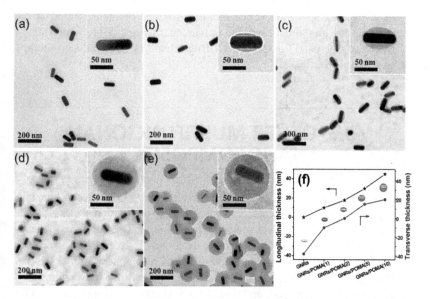

FIGURE 3.3 TEM images of (a) gold nanorod, (b–e) gold nanorod coated with conducting polymer of increased coating thickness, and (f) the longitudinal and transverse coating thickness. Reprinted with permission from Wang, J. et al. 2018. Controllable synthesis of gold nanorod/conducting polymer core/shell hybrids toward in vitro and in vivo near-infrared photothermal therapy. *ACS Applied Materials & Interfaces*, 10, 12323–12330. Copyright 2018 American Chemical Society.

and the introduction of additional properties. Among them, coating and functionalization with small molecules for biological targeting is discussed in detail in Chapter 4. Here we will briefly discuss how functionalization with small molecules introduces additional properties along with some selected examples.

In one example, polymer-coated gold nanorod is conjugated with fluorescent dye to produce plasmonic-fluorescent nanorods.[25] This material can be used for dual imaging nanoprobes. In this approach, a separation distance of >5 nm between the nanorod and dye is important to minimize the fluorescence quenching of the dye by the gold nanorod. In another approach, a dye-conjugated gold nanorod is prepared to demonstrate the coupling of nanorod plasmon with excited state of dye.[26] In another case, Raman active dye-conjugated nanorod is prepared for application as a surface enhanced Raman spectroscopy (SERS) probe.[27] Here, the surface plasmon of the nanorod is used to enhance the Raman signal of the dye. In both cases, the dyes are spaced close to the nanorod surface in order to produce the expected result.

The photoresponsive property of gold nanorods is used for NIR light-responsive drug release and to modulate drug performance.[28] In these cases, drugs are adsorbed at the nanorod surface or incorporated at the shell structure and then used for enhanced therapy. More importantly, NIR light can be used for responsive drug release or light-induced therapy applications.

3.6 OTHER MULTIFUNCIONAL NANOCOMPOSITES

The unique properties of gold nanorods have inspired the preparation of a variety of multifunctional nanocomposites. Here, we will briefly discuss some of them with examples. Gold nanorod, graphene, and hemin-based ternary composites are designed for the sensitive detection of glycan that is expressed on specific cell surfaces.[29] The nanorod–graphene component mimics the peroxidase-like property. The nanocomposite is functionalized for specific labeling of glycan-expressed cells and then the peroxidase-like property is used for the detection of glycan.

In another example, gold nanorod and calcium phosphate-based composites are prepared using a polymer assembly approach.[30] This material offers high drug loading, dual-responsive drug release, and photothermal properties. This material offers better therapeutic effects than single chemotherapy or photothermal therapy. In another approach, gold nanorod-conjugated porous silicon is prepared and encapsulated in calcium alginate hydrogel for preparing nanohydrogel.[31] This material is shown to have controlled drug-release properties and photothermal effects.

In another example, a short wave infrared photodetector is designed via composite formation between gold nanorod and platinum microwire.[32] Platinum microwires harvest the photothermal effects of gold nanorods and generate a change in device resistance. A fast photon-to-heat-to-resistance conversion using this composite-based photodetector offers microsecond response times.

REFERENCES

1. Hana, C., Qi, M.-Y., Tanga, Z.-R., Gongc, J. and Xu, Y.-J. 2019. Gold nanorods-based hybrids with tailored structures for photoredox catalysis: Fundamental science, materials design and applications. *Nano Today*, 27, 48–72.

2. Shukla, N., Singh, B., Kim, H.-J., Park, M.-H and Kim, K. 2020. Combinational chemotherapy and photothermal therapy using a gold nanorod platform for cancer treatment. *Particle & Particle Systems Characterization*, 37, 2000099.

3. Zheng, J., Cheng, X., Zhang, H., Bai, X., Ai, R., Shao, L. and Wang, J. 2021. Gold nanorods: The most versatile plasmonic nanoparticles. *Chemical Reviews*, 121, 13342–13453.

4. Focsan, M., Gabudean, A. M., Vulpoi, A. and Astilean, S. 2014. Controlling the luminescence of carboxyl-functionalized CdSe/ZnS core–shell quantum dots in solution by binding with gold nanorods. *The Journal of Physical Chemistry C*, 118, 25190–25199.

5. Rittikulsittichai, S., Kolhatkar, A. G., Sarangi, S., Vorontsova, M. A., Vekilov, P. G., Brazdeikis A. and Lee, T. R. 2016. Multi-responsive hybrid particles: thermo-, pH-, photo-, and magneto-responsive magnetic hydrogel cores with gold nanorod optical triggers. *Nanoscale*, 8, 11851–11861.

6. Wu, B., Liu, D., Mubeen, S., Chuong, T. T., Moskovits, M. and Stucky, G. D. 2016. Anisotropic growth of TiO_2 onto gold nanorods for plasmon-enhanced hydrogen production from water reduction. *Journal of the American Chemical Society*, 138, 1114–1117.

7. Chakraborty, A., Fernandez, A. C., Som, A., Mondal, B., Natarajan, G., Paramasivam, G., Lahtinen, T., Nonappa, H. H. and Pradeep, T. 2018. Atomically precise nanocluster assemblies encapsulating plasmonic gold nanorods. *Angewandte Chemie International Edition*, 57, 6522–6526.

8. Chan, M.-H., Chen, S.-P., Chen, C.-H., Chan, Y.-C., Lin, R. J., Tsai, D. P., Hsiao, M., Chung, R.-J., Chen, X. and Liu, R.-S. 2018. Single 808 nm laser treatment comprising photothermal and photodynamic therapies by using gold nanorods hybrid upconversion particles. *The Journal of Physical Chemistry C*, 122, 2402–2412.

9. He, L., Mao, C., Brasino, M., Harguindey, A., Park, W., Goodwin, A. P. and Cha, J. N. 2018. TiO_2-capped gold nanorods for plasmon-enhanced production of reactive oxygen species and photothermal delivery of chemotherapeutic agents. *ACS Applied Materials & Interfaces*, 10, 27965–27971.

10. Pavelka, O., Dyakov, S., Veselý, J., Fučíková, A., Sugimoto, H., Fujii, M. and Valenta, J. 2021. Optimizing plasmon enhanced luminescence in silicon nanocrystals by gold nanorods. *Nanoscale*, 13, 5045–5057.

11. Wang, J.-H., Chen, M., Luo, Z.-J., Ma, L., Zhang, Y.-F., Chen, K., Zhou, L. and Wang, Q.-Q. 2016. Ceria-coated gold nanorods for plasmon-enhanced near-infrared photocatalytic and photoelectrochemical performances. *The Journal of Physical Chemistry C*, 120, 14805–14812.

12. Wang, B., Li, R., Guo G. and Xia, Y. 2020. Janus and core@shell gold nanorod@$Cu_{2-x}S$ supraparticles: reactive site regulation fabrication, optical/catalytic synergetic effects and enhanced photothermal efficiency/photostability. *Chemical Communications*, 56, 8996–8999.

13. Aioub, M., Panikkanvalappil, S. R. and El-Sayed, M. A. 2017. Platinum-coated gold nanorods: Efficient reactive oxygen scavengers that prevent oxidative damage toward healthy, untreated cells during plasmonic photothermal therapy. *ACS Nano*, 11, 579–586.

14. Sun, L., Zhang, Q., Li, G. G., Villarreal, E., Fu, X. and Wang, H. 2017. Multifaceted gold–palladium bimetallic nanorods and their geometric, compositional, and catalytic tunabilities. *ACS Nano*, 11, 3213–3228.

15. Li, H., Chen, J., Fan, H., Cai, R., Gao, X., Meng, D., Ji, Y., Chen, C., Wang, L. and Wu, X. 2020. Initiation of protective autophagy in hepatocytes by gold nanorod core/silver shell nanostructures. *Nanoscale*, 12, 6429–6437.

16. Ortega-Liebana, M. C., Hueso, J. L., Arenalc, R. and Santamaria, J. 2017. Titania-coated gold nanorods with expanded photocatalytic response. Enzyme-like glucose oxidation under near-infrared illumination. *Nanoscale*, 9, 1787–1792.

17. Guo, T., Lin, Y., Li, Z., Chen, S., Huang, G., Lin, H., Wang, J., Liu, G. and Yang, H.-H. 2017. Gadolinium oxysulfide-coated gold nanorods with improved stability and dual-modal magnetic resonance/photoacoustic imaging contrast enhancement for cancer theranostics. *Nanoscale*, 9, 56–61.

18. Ma, K., Li, Y., Wang, Z., Chen, Y., Zhang, X., Chen, C., Yu, H., Huang, J., Yang, Z., Wang, X. and Wang, Z. 2019. Core–shell gold nanorod@layered double hydroxide nanomaterial with highly efficient photothermal conversion and its application in antibacterial and tumor therapy. *ACS Applied Materials & Interfaces*, 11, 29630–29640.

19. Zhang, L., Su, H., Cai, J., Cheng, D., Ma, Y., Zhang, J., Zhou, C., Liu, S., Shi, H., Zhang, Y. and Zhang, C. 2016. A multifunctional platform for tumor angiogenesis-targeted chemo-thermal therapy using polydopamine-coated gold nanorods. *ACS Nano*, 10, 10404–10417.

20. Liu, S., Wang, L., Lin, M., Wang, D., Song, Z., Li, S., Ge, R., Zhang, X., Liu, Y., Li, Z., Sun, H., Yang, B. and Zhang, H. 2017. Cu(II)-doped polydopamine-coated gold nanorods for tumor theranostics. *ACS Applied Materials & Interfaces*, 9, 44293–44306.

21. Osterrieth, J. W. M., Wright, D., Noh, H., Kung, C.-W., Vulpe, D., Li, A., Park, J. E., Van Duyne, R. P., Moghadam, P. Z., Baumberg, J. J., Farha, O. K. and Fairen-Jimenez, D. 2019. Core–shell gold nanorod@zirconium-based metal–organic framework composites as in situ size-selective Raman probes. *Journal of the American Chemical Society*, 141, 3893–3900.

22. Zhou, Z., Zhao, J., Di, Z., Liu, B., Li, Z., Wu, X. and Li, L. 2021. Core–shell gold nanorod@mesoporous-MOF heterostructures for combinational phototherapy. *Nanoscale*, 13, 131–137.

23. Reiser, B., Gonzalez-Garcıa, L., Kanelidis, I., Maurer, J. H. M. and Kraus, T. 2016. Gold nanorods with conjugated polymer ligands: Sintering-free conductive inks for printed electronics. *Chemical Science*, 7, 4190–4196.

24. Wang, J., Zhu, C., Han, J., Han, N., Xi, J., Fan, L. and Guo, R. 2018. Controllable synthesis of gold nanorod/conducting polymer core/shell hybrids toward in vitro and in vivo near-infrared photothermal therapy. *ACS Applied Materials & Interfaces*, 10, 12323–12330.

25. Basiruddin, Sk., Saha, A., Pradhan, N. and Jana, N. R. 2010. Functionalized gold nanorod solution via reverse micelle based polyacrylate coating. *Langmuir*, 26, 7475–7481.

26. Hao, Y.-W., HWang, H.-Y., Jiang, Y., Chen, Q.-D., Ueno, K., Wang, W.-Q., Misawa, H., and Sun, H.-B. 2011. Hybrid-state dynamics of gold nanorods/dye J-aggregates under strong coupling. *Angewandte Chemie International Edition*, 50, 7824–7828.

27. Wang, Y., Wang, Y. Q., Wang, W., Sun, K. and Chen, L. 2016. Reporter-embedded SERS tags from gold nanorod seeds: Selective immobilization of reporter molecules at the tip of nanorods. *ACS Applied Materials & Interfaces*, 8, 28105–28115.

28. Bhana, S., O'Connor, R., Johnson, J., Ziebarth, J. D., Henderson, L. and Huang, X. 2016. Photosensitizer-loaded gold nanorods for near infrared photodynamic and photothermal cancer therapy. *Journal of Colloid and Interface Science*, 469, 8–16.
29. Liu, J., Xin, X., Zhou, H. and Zhang, S. 2015. A ternary composite based on graphene, hemin, and gold nanorods with high catalytic activity for the detection of cell-surface glycan expression. *Chemistry – A European Journal*, 21, 1908–1914.
30. Li, S., Zhang, L., Zhang, H., Mu, Z., Li, L. and Wang, C. 2017. Rationally designed calcium phosphate/small gold nanorod assemblies using poly(acrylic acid calcium salt) nanospheres as templates for chemo-photothermal combined cancer therapy. *ACS Biomaterials Science & Engineering*, 3, 3215–3221.
31. Zhang, H., Zhu, Y., Qu, L., Wu, H., Kong, H., Yang, Z., Chen, D., Makila, E., Salonen, J., Santos, H. A., Hai, M. and Weitz, D. A. 2018. Gold nanorods conjugated porous silicon nanoparticles encapsulated in calcium alginate nano hydrogels using microemulsion templates. *Nano Letters*, 18, 1448–1453.
32. Xiang, H., Hu, Z., Billot, L., Aigouy, L. and Chen, Z. 2019. Hybrid plasmonic gold-nanorod–platinum short-wave infrared photodetectors with fast response. *Nanoscale*, 11, 18124–18131.

Surface Chemistry and Functionalization of Gold Nanorod

4

4.1 WHY SURFACE CHEMISTRY AND FUNCTIONALIZATION?

As synthesized gold nanorod is coated with a surfactant (cetyltrimethyl ammonium bromide, CTAB) double layer. This surfactant coating makes nanorods cationic in nature. In addition, the lipophilic nature of CTAB makes the nanorod lipophilic. This CTAB coating offers water-dispersible nanorods with high colloidal stability. However, CTAB is physically adsorbed at the surface of the nanorod and it can be removed by repeated precipitation (via centrifuge) and re-dispersion in fresh water. These processes will lead to insoluble and aggregated nanorods. Moreover, different applications require nanorods with specific surface chemistry. For example, cell targeting or sub-cellular targeting requires specific biomolecule termination, longer stability in blood circulation requires polyethylene glycol termination, and charged nanorods are required for electrostatic-interaction-based processing. Thus surface modification of synthesized nanorods by various chemicals/biochemicals is critical for further application.[1,2] In this chapter, we will discuss different approaches in deriving varied surface chemistry and in making various functional gold nanorods. Table 4.1 summarizes different approaches that have been developed and will be discussed here.[3–38] They include the exchange of CTAB by thiolated small molecules or polymers,[3–19] binding of anionic molecules/polymers with cationic CTAB at the surface of the nanorod via electrostatic interaction,[20–28] encapsulation of nanorods inside polymer micelle,[29] and replacement of the CTAB layer by silica shell[30–33] or polymer multilayers.[34–38] In addition, the

DOI: 10.1201/9781003245339-4

TABLE 4.1 Summary of Coating Chemistry for Gold Nanorods along with Their Application Potential.

METHOD	CHEMICALS USED	COATING TYPE AND THICKNESS	APPLICATION	REFERENCE
Ligand exchange	thiolated peptide	molecular monolayer	nuclear targeting	3–5
Ligand exchange	thiolated oligonucleotide	molecular monolayer	cell uptake, gene delivery	6,7
Ligand exchange	mercaptoalkanoic acid	molecular monolayer	biological functionalization	8,9
Ligand exchange	N-heterocyclic-carbene–thiolate	molecular monolayer	photothermal therapy	10
Ligand exchange	thiolated lactose	molecular monolayer	biosensor	11
Ligand exchange	thiolated chitosan oligosaccharide	molecular monolayer	bioconjugation	12
Ligand exchange	thiolated dextran	dextran monolayer	biocompatibility	13
Ligand exchange	thiolated glycan	molecular monolayer	enhanced colloidal stability	14
Ligand exchange	thiolated polyethylene glycol	molecular monolayer	enhanced biocompatibility, therapy	15,16
Ligand exchange	thiolated poly(ethylene oxide)	polymer monolayer	enhanced blood circulation	17
Ligand exchange	thiolated poly(N-isopropylacrylamide)	polymer encapsulation	light-responsive drug delivery	18
Ligand exchange	thiolated polyvinyl alcohol	polymer monolayer	enhanced biocompatibility	19
Ligand exchange	phospholipid	molecular mono or bilayer	lowering of cytotoxicity	20,21
Ligand exchange	citrate	molecular monolayer	biocompatibility, functionalization	22

(Continued)

TABLE 4.1 (CONTINUED)   Summary of Coating Chemistry for Gold Nanorods along with Their Application Potential.

METHOD	CHEMICALS USED	COATING TYPE AND THICKNESS	APPLICATION	REFERENCE
Electrostatic binding	ethylenediamine tetaacetic acid	molecular monolayer	biosensor	23
Electrostatic binding	polyacrylic acid	polymer monolayer	peptide conjugation, cell therapy	24
Protein coating	serum protein	protein monolayer	enhanced biocompatibility, light-responsive drug delivery	25–28
Encapsulation in micelle	block copolymer	polymer monolayer	enhanced colloidal stability	29
Silica coating	silane	cross-linked layer of 2–50 nm	biocompatibility, functionalization	30–33
Layer-by-layer coating	cationic and anionic polymers	multilayers of 1–10 nm	peptide conjugation, cell targeting	34, 35
Polyacrylate coating	acryl monomers	cross-linked polymer shell of 5–10 nm	bioconjugation chemistry	36
Carbon coating	heating with glucose	cross-linked multilayers of 10–25 nm	biocompatibility, enhanced photothermal therapy	37
Metal–organic framework coating	1,3,5-benzenetricarboxylate, Cu^{+2}	cross-linked multilayers of 2–10 nm	multifunctional	38

coating materials are further designed in a way that they are terminated with specific chemicals/biochemicals or can be used for covalent conjugation of other molecules.

4.2 THIOL-BASED LIGAND EXCHANGE FOR COATING AND FUNCTIONALIZATION

Thiol-based small molecules and polymers are most commonly used for replacing CTAB from gold nanorod surfaces.[3–19] This method is commonly known as the ligand exchange approach. Figure 4.1 shows a schematic representation of thiol-based ligand exchange. Thiols are selected because sulfur atoms of the thiol group have a strong affinity to gold atoms that induce efficient chemisorption. The thiolated small molecules that are used include alkanoic acids, peptides, oligonucleotides, biopolymers, synthetic polymers, and polyethylene glycols. In addition, specific biomolecules are converted into their thiolated form so that they are chemisorbed at the nanorod surface through the thiol group and produce biomolecule-terminated nanorods. Polymers with multiple thiols or small molecules with multiple thiols have an additional advantage and they can anchor more efficiently on the nanorod surface via multiple thiol groups. However, the ligand exchange condition should be adjusted carefully to minimize the nanorod–nanorod crosslinking.

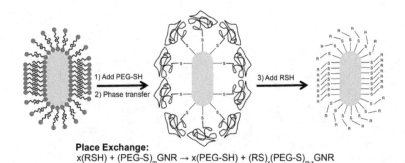

Place Exchange:
x(RSH) + (PEG-S)$_m$GNR → x(PEG-SH) + (RS)$_x$(PEG-S)$_{m-x}$GNR

FIGURE 4.1 Schematic representation of the exchange of CTAB at the gold nanorod surface by thiolated PEG followed by an exchange with other alkanethiols. Reprinted with permission from Burrows, N. D. et al. 2016. Surface chemistry of gold nanorods. *Langmuir*, 32, 9905–9921. Copyright 2016 American Chemical Society.

There are two common issues on thiol-based ligand exchange. First, the ligand exchange is often incomplete and only the CTAB molecules at either end of nanorods are completely replaced, while many CTAB molecules on the long side still remain. This is particularly due to the effective capping of CTAB at side faces than at end faces. Second, nanorods aggregate during ligand exchange. This is due to the loosening of CTAB capping and charge reversal issues during ligand exchange steps. Thus various approaches have been developed for efficient ligand exchange. Here we will briefly highlight those approaches.

In one approach, the CTAB bilayer around the gold nanorod is replaced by 11-mercaptoundecanoic acid via a biphasic mixture of water–ionic liquid.[8] Here ionic liquid assists efficient ligand exchange without nanorod aggregation. 11-Mercaptoundecanoic acid is frequently used to replace CTAB, as the thiol group strongly binds to the Au surface and the carboxylic acid group can be used for conjugation with biomolecules. In another approach, excess CTAB is removed prior to ligand exchange but without any nanorod aggregation.[9] This has been achieved by precipitating the nanorods via centrifuge and then re-dispersing them in dilute CTAB solution. This process can be repeated several times prior to adding mercaptoalkanoic acid for ligand exchange.

Similarly, one-step surface functionalization with thiolated polyethylene glycol (PEG) has been demonstrated in the presence of Tween 20, bis(p-sulfonatophenyl)phenylphosphine, and NaCl.[15] Here Tween 80 offers stabilization of the nanorod during ligand exchange, bis(p-sulfonatophenyl) phenylphosphine offers activation of the nanorod surface for PEGylation, and NaCl offers etching of silver from the nanorod surface. This method allows for the complete removal of CTAB from the nanorod surface. In another approach. 11-mercaptoundecanoic acid conjugated PEG is used to replace CTAB layers around nanorods. This leads to more effective PEGylation of nanorods with enhanced colloidal stability.[16]

A wide variety of peptides are known for enhanced cell uptake and subcellular targeting. These peptides are transformed into thiolated peptides or cysteine-terminated peptides and are used for ligand exchange. For example, gold nanorods are functionalized with nuclear localization signal peptides via thiol-based modification of the peptide.[3] These functional nanorods are used for targeting the cell nucleus. Similarly, 11-mercaptoundecyl phosphorylcholine and cysteine-terminated TAT peptides are used separately for the replacement of CTAB from the nanoparticle surface via the ligand exchange approach.[4,5] Resultant nanorods with phosphorylcholine or TAT are used for NIR light-induced cancer cell therapy. In an interesting approach, a bidentate thiolate–N-heterocyclic-carbene–gold (I) is designed and exchanged with CTAB at the gold nanorod surface.[10] Next, a mild reduction condition is used that converts gold (I) to gold adatom. The resulting thiolated–heterocyclic-carbene-stabilized gold nanorods are stable towards excess glutathione and under conditions in biological media and cell culture.

It is a difficult challenge to attach thiolated oligonucleotides on gold nanorods as CTAB-capped positively charged nanorods readily aggregate in the presence of negatively charged oligos. A method was developed that uses thiolated PEG and surfactant (Tween 20)-assisted capping of the thiolated oligonucleotide.[6] Tween 20 and thiolated PEG synergistically displace CTAB from the nanorod surface via repeated centrifugation and resuspension. Next, thiolated oligos are attached to the nanorod surface in the presence of salt/citrate. The number of loaded oligo per nanorod can be controlled by varying the concentrations of thiolated PEG or the ratio between oligo and nanorod. Functionalized nanorods are used for nanoparticle assembly and cancer cell imaging. Similarly, the thiolated oligonucleotide is used for the capping of nanorods via the exchange of CTAB. This nanorod is then used for length effect in cell uptake mechanisms.[7]

Several polymers are transformed into thiol-based polymers for capping with gold nanorods. For example, thiolated dextran is synthesized via covalent linking of dextran with mercaptopropionic acid.[13] Next, thiol-functionalized dextran solution is mixed with CTAB-coated gold nanorod. After 24 h, the dextran-coated nanorod is isolated via centrifuge, and supernatants with unbound dextran are removed. Finally, dextran-coated nanorods are resuspended in fresh water. This material showed enhanced bio-compatibility. Similarly, gold nanorods coated with thiolated polyvinyl alcohol are prepared via a two-step route.[19] First, CTAB-capped nanorod is exposed with thiolated polyvinyl acetate in dimethyl formamide. At this stage, CTAB present at nanorod tips are replaced by thiolated polyvinyl acetate. Next, nanorods are again exposed with thiolated polyvinyl acetate in methanol/ethanol that replaces the rest of the CTAB using thiolated polyvinyl acetate. Finally, exposure to a basic medium leads to the formation of thiolated polyvinyl alcohol-terminated nanorods. These nanorods have demonstrated good bio-compatibility and colloidal stability. Gold nanorods capped with thiolated glycan are prepared via ligand exchange of CTAB by thiolated glycan.[14] Glycan-coated nanorods are resistant to the adsorption of proteins from serum-containing media and avoid phagocytosis by macrophage-like cells, but retain selectivity toward carbohydrate-binding proteins. This work shows that glycan can be an alternative to PEG in inhibiting non-specific interaction. The grafting of poly(ethylene oxide) (PEO) onto the nanorod surface is achieved via the exchange of CTAB by thiolated poly(ethylene oxide).[17] This process minimizes non-specific interaction and offers longer circulation of nanorods in blood. In another approach, gold nanorod is capped with temperature-responsive polymer poly(N-isopropylacrylamide).[18] Here CTAB at the gold nanorod surface is replaced by thiolated poly(N-isopropylacrylamide).

The resultant coated nanorod is loaded with drug and used for responsive drug delivery. Under NIR light exposure, the nanorod induces local heating that makes the polymer hydrophobic and thus releases the drug.

4.3 ELECTROSTATIC AND OTHER INTERACTION-BASED COATING AND FUNCTIONALIZATION

As CTAB-capped gold nanorods are cationic in nature, any anionic molecules/polymers can be coated/attached via electrostatic attraction. In addition, the lipophilic character of attached CTAB can offer attachment of hydrophobic groups of molecule/polymer via hydrophobic interaction. These aspects are used for the coating and functionalization of nanorods. For example, CTAB-capped gold nanorods are transformed into phospholipid-coated nanorods via ligand exchange.[20,21] Typically, nanorods are precipitated by centrifuge, and then they are re-suspended in phospholipid solution along with bath sonication. The centrifugation and sonication are then repeated several times until CTAB is completely removed. In another example, citrate-stabilized gold nanorods are prepared via a two-step approach.[22] At first, CTAB at the nanorod surface is exchanged with polystyrenesulfonate. Next, polystyrenesulfonate is replaced by citrate. Typically, gold nanorod dispersion is subjected to centrifuge to precipitate the nanorods. Next, nanorods are re-dispersed in 0.15 wt % polystyrenesulfonate solution. This type of centrifugation and re-dispersion in polystyrenesulfonate solution is repeated 3–4 times and that completely replaces CTAB with polystyrenesulfonate. Next, polystyrenesulfonate-coated nanorods are precipitated by centrifuge and re-dispersed in 5 mM sodium citrate for the exchange of polystyrenesulfonate by citrate. This type of centrifugation and re-dispersion in citrate solution is repeated 2–3 times and that completely replaces polystyrenesulfonate with citrate. Similarly, reversible assembly and disassembly of gold nanorods have been achieved via electrostatic interaction between CTAB bilayer present on the gold nanorod surface and anionic ethylenediamine tetra-acetic acid (EDTA).[23] The EDTA concentration has been adjusted for nanorod assembly via lowering of surface charge. These assemblies can be broken by adding metal ions via complexation with EDTA. This approach has been used for the optical detection of metal ions.

Protein coating has a great appeal for making nanorods biocompatible. Thus different approaches that utilize electrostatic and hydrophobic interaction have been developed for protein coating.[25–28] In addition, thiol-based peptides may be present as protein components that further enhance the binding with nanorods. For example, the ligand exchange approach has been used for replacing CTAB at gold nanorod surface using bovine serum albumin.[27] It has been shown that protein-coated nanorods show enhanced colloidal stability and biocompatibility. In a typical procedure, CTAB-stabilized nanorod dispersions are prepared by removing excess CTAB. Next, nanorod dispersions are added slowly to the protein solution under ultra sonication. Then nanorods are isolated by centrifuge and re-dispersed in fresh protein solution. Finally, nanorods are isolated via centrifugation and concentrated as desired. In another report, protein-coated nanorods are used for cell therapy via mitochondria targeting.[25] Moreover, these types of protein-coated nanorods are used as drug delivery carriers.[26,28] In particular, multiple hydrophobic/hydrophilic/charged drugs are loaded at protein corona and used for multimodal cancer therapy. In addition, light can be used for controlled drug release and photothermal therapy.

In an alternative approach, the self-assembly processing technique has been exploited for encapsulating gold nanorods using a polymer. Block copolymer micelle is used for encapsulating gold nanorods.[29] In this process CTAB-capped nanorods are mixed with block copolymer in water-dimethylformamide solvent and under this condition, polymers self-assemble at the nanorod surface with the formation of a dense polymer brush layer. The resultant nanorod has high colloidal stability even under high ionic strength conditions.

4.4 SILICA COATING APPROACH

Silica coating of gold nanorods appears as an attractive alternative for their manipulation and for providing enhanced colloidal stability of gold nano rod.[1,2,30–33] Figure 4.2 shows a typical electron microscopic image of gold nanorod coated with varied silica shell thickness. The coating method involves the replacement of CTAB at the nanorod surface by silane/silicate precursor followed by silica shell formation around the nanorod. However, CTAB strongly binds to the gold nanorod surface which makes it difficult to displace silane/silicate precursors. The most effective silane precursor is mercapto-silane. This is because it has a mercapto group that chemisorbs to the nanorod surface and a silane component acts as a nucleation site for silica formation. Another issue of silica coating is that nanorods are dispersible in water but most of the silane precursors are insoluble in water and require polar organic solvent to solubilize them. Thus silica coating conditions need to be

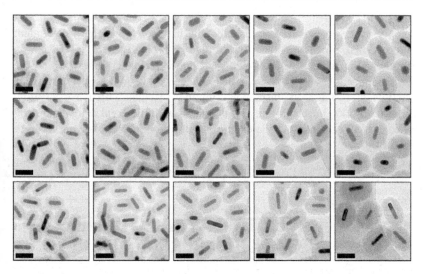

FIGURE 4.2 TEM image of gold nanorod coated with varied silica shell thickness.

carefully adjusted so that both nanorod and silica shell forming precursors are dispersed/soluble in reaction media.

In the earliest approach of silica coating, the CTAB-capped gold nanorod is mixed with (3-mercaptopropyl)trimethoxysilane (MPTMS) solution. At this stage, the exchange of CTAB occurs by MPTMS and acts as a silica nucleation site. Next, freshly prepared aqueous sodium silicate is added as a source of silica. This leads to the formation of a silica shell around the nanorod.[30] However, this approach leads to a significant percentage of aggregated nanorods.

In later stages, alcohol–water mixture is used for homogeneous silica coating via hydrolysis and condensation of tetraethoxy silane.[31] In this approach, CTAB at the gold nanorod surface is replaced by polymer so that nanorods can be dispersed in the alcohol–water mixture. Next, subsequent silica coating is performed through the well-known Stöber method where tetraethoxy silane hydrolyze and condenses as a silica shell. Following this approach, silica-coated gold nanorods with controlled shell thickness in the range of 1–50 nm can be achieved. The resultant silica-coated nanorods can be dispersed in water and other organic solvents.

Following these works, a variety of silica shell growth conditions are reported to vary the shell thickness from a few nanometers to 100 nm, controlling the porosity of silica shell and silica selling at either end or side of rods.[1,33] The thin shells are suitable for applications such as plasmonic biosensing, while a thicker and porous shell is suited for drug encapsulation and controlled drug release. For example, silica-coated gold nanorod has been used to investigate the distance and plasmon wavelength-dependent fluorescence properties of infrared dye.[32] Typically, dye is covalently bound to the surface of the silica shell and it has been observed that the fluorescence intensity of dye can be enhanced by 10-fold for a 17 nm silica shell when the nanorod plasmon maximum is resonant with dye absorption. In other works, porous silica cell is loaded with different drugs and drug release has been induced via excitation of gold nanorod plasmon by light exposure.[1]

4.5 LAYER-BY-LAYER POLYMER COATING APPROACH

Laye-by-layer polymer coating is an attractive approach for coating and functionalization of gold nanorods.[34] This approach utilizes alternate adsorption of anionic and cationic polyelectrolytes that leads to the formation of polyelectrolyte multilayers around the nanorods. At first, CTAB-capped nanorod with cationic surface charge is used for adsorption of anionic polymer via electrostatic attraction. The resultant anionic nanorods are then used for adsorption of cationic polymer. In the first report, poly(sodium-4-styrenesulfonate) was used as anionic polymer and poly(diallyldimethylammonium chloride) was used as cationic polymer (see Figure 4.3). One key issue is the nanorod aggregation during the coating steps. This is due to the multiple cationic/anionic charges present both in polymer and coated nanorods that can cross-link nanorods. Such cross-linking can be minimized using dilute solutions or using salts that screen some of the charge.

In a typical process, stock solutions of cationic/anionic polymers are prepared separately in aqueous NaCl solution. Next, synthesized gold nanorods are precipitated via centrifuge to remove excess CTAB. Then CTAB-capped nanorods are dispersed in anionic polymer solution. Next, excess polymer is removed via precipitating nanorods by centrifuge and discarding the supernatant fraction. Precipitated nanorods are then dispersed in a cationic polymer solution. Successive repetition of this step has been used to form polymer multilayers. The coated nanorods have colloidal properties with surface charge depending on the polymer used in the last step. The coating thickness can be varied from 1 to 10 nm by varying multilayers and nanorods can be further functionalized via conjugation chemistry using functional groups at the surface or via electrostatic binding.[35]

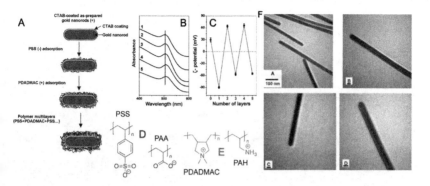

FIGURE 4.3 A) Schematic diagram illustrating the layer-by-layer coating process around gold nanorod. B, C) Visible spectra and Zeta potential data of gold nanorods at successive coating steps. D, E) Common anionic and cationic polymers used for layer-by-layer assembly. F) TEM image of gold nanorods at different stages of coating. (A–E) Reprinted with permission from Burrows, N. D. et al. 2016. Surface chemistry of gold nanorods. *Langmuir*, 32, 9905–9921. Copyright 2016 American Chemical Society. (F) Reprinted with permission from Gole, A. et al. 2005. Polyelectrolyte-coated gold nanorods: Synthesis, characterization and immobilization. *Chemistry of Materials*, 17, 1325–1330. Copyright 2005 American Chemical Society.

4.6 OTHER COATINGS WITH CROSS-LINKED SHELL

Other coating methods have been developed that offer good colloidal stability, options for conjugation chemistry, and enhanced photothermal performance of gold nanorods. In particular, various approaches are designed for robust coating with cross-linked shells that ensure better protection of nanorods.[36–38]

In one example, polyacrylate coating has been developed around gold nanorods and offers options for bioconjugation chemistry at the nanorod surface with good colloidal stability of nanobioconjugates[36] (Figure 4.4). A key aspect of this approach is the in situ polymerization at the nanorod surface and cross-linked coating of 5 to 20 nm thickness, terminated with a wide variety of chemical functional groups. The polymerization has been performed in reverse micelle so that nanorods and acryl monomers that are hydrophilic/hydrophobic in nature can be dispersed/solubilized in a single phase. This approach produces water-dispersible nanorods terminated with primary amine, ammonium, sulfonate, polyethylene glycol, and other functional groups. In addition, a wide variety of conjugation chemistry can be exercised for making various nanobioconjugates.[36]

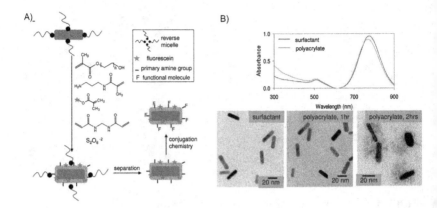

FIGURE 4.4 A) Reverse micelle-based polyacrylate coating and functionalization scheme for gold nanorods. B) UV–visible absorption spectra (top panel) and TEM images (bottom panel) of gold nanorods before and after polyacrylate coating. Reprinted with permission from Basiruddin, S. K. et al. 2010. Functionalized gold nanorod solution via reverse micelle based polyacrylate coating. *Langmuir,* 26, 7475–7481. Copyright 2010 American Chemical Society.

In another approach, carbon-coated gold nanorod has been designed with good biocompatibility and high-performance photothermal therapy. Colloidal gold nanorod is mixed with glucose and heated at 180 °C in an autoclave for generating a carbon shell around the nanorod. The carbon shell thicknesses have been controlled from 10 to 25 nm by varying reaction conditions. These nanorods show enhanced photothermal performance and low cytotoxicity.[37]

In another approach, a gold nanorod with a metal–organic framework shell was designed to achieve multifunctional properties.[38] The approach involves layer-by-layer deposition of Cu^{+2} and 1,3,5-benzenetricarboxylate on the colloidal solution of mercaptoundecanoic acid-capped nanorod (Figure 4.5). The resultant coating is cross-linked in nature and coating thickness can be controlled between 2 and 10 nm depending on the number of multilayers. Resultant nanorods are colloidal in nature and the metal–organic framework shell offers additional properties.

4.7 FUNCTIONALIZATION OF GOLD NANOROD VIA DIFFERENT APPROACHES

Coating and surface chemistry exercises on gold nanorod surfaces are most important for their functionalization. There are three common approaches

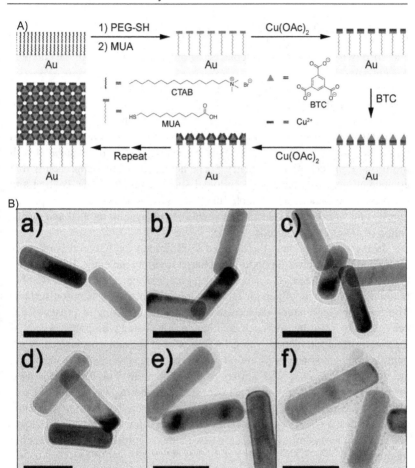

FIGURE 4.5 A) Schematic representation of making metal–organic framework shell around gold nanorod. B) Transmission electron microscopy images of coated gold nanorods with increasing layers. Scale bars are 50 nm. Reprinted with permission from Hinman, J. G. et al. 2018. Layer-by-layer synthesis of conformal metal–organic framework shells on gold nanorods. *Chemistry of Materials*, 30, 7255–7261. Copyright 2018 American Chemical Society.

for functionalization. In the first approach, thiolated affinity molecule is prepared separately and then they are used to exchange CTAB at the nanorod surface. This approach is commonly used to make small molecule-terminated nanorods. Examples include peptide-functionalized nanorods,[3–5] oligonucleotide-functionalized nanorods,[6,7] and lactose-functionalized nanorods.[11] Small molecules are used because the preparation and characterization of thiolated small molecule and their ligand exchange are relatively simple. However, the colloidal property of functional nanorods is poor in many cases and affinity molecules may be replaced by other thiols present in the complex bio-environment.

In the second approach, a nanorod with a robust coating is used for conjugation with affinity molecules. Here, the chemical functional group at the coated nanorod surface is used for conjugation with the functional group of the affinity molecule. For example, polyacrylate coated nanorod is conjugated with glucosamine using glutaraldehyde-based conjugation or conjugated with biotin via EDC coupling chemistry.[36] The advantage of this approach is that conjugated nanorods have good colloidal stability due to robust coating and the presence of other water-soluble functional groups. However, this approach requires the appropriate selection of chemical conjugation method and purification of conjugated nanorods.

In the third approach, the charged shell around the nanorod is used for electrostatic assembly with oppositely charged affinity molecule. For example, peptide-functionalized gold nanorods are prepared via electrostatic assembly of anionic component of peptide with cationic polymer at the nanorod surface. The functionalized nanorods are used for specific targeting of prostate cancer cells.[35] In a similar approach gold nanorod is coated with polyacrylic acid via electrostatic interaction with CTAB present at the nanorod surface. Next, peptide-conjugated nanorods are prepared via conjugation between carboxylate groups of nanorods with amine groups of peptide. This material is used for photothermal therapy of breast cancer.[24] Considering the wide variety of affinity molecules/biochemicals and that each of them have different chemical properties, a variety of approaches are required for their functionalization and this field will continue to develop new methods or modification of existing methods.[39,40]

REFERENCES

1. Burrows, N. D., Lin, W., Hinman, J. G., Dennison, J. M., Vartanian, A. M., Abadeer, N. S., Grzincic, E. M., Jacob, L. M., Li, J. and Murphy, C. J. 2016. Surface chemistry of gold nanorods. *Langmuir*, 32, 9905–9921.
2. Jana, N. R. *Colloidal Nanoparticles: Functionalization for Biomedical Applications*. CRC Press, Boca Raton, 2019.
3. Oyelere, A. K., Chen, P. C., Huang, X., El-Sayed, I. H. and El-Sayed, M. A. 2007. Peptide-conjugated gold nanorods for nuclear targeting. *Bioconjugate Chemistry*, 18, 1490–1497.

4. Zhou, W., Liu, X. and Ji, J. 2012. Fast and selective cancer cell uptake of therapeutic gold nanorods by surface modifications with phosphorylcholine and Tat. *Journal of Materials Chemistry*, 22, 13969–13976.

5. Bartneck, M., Ritz, T., Keul, H. A., Wambach, M., Bornemann, J., Gbureck, U., JEhling, J., Lammers, T., Heymann, F., Gassler, N., Lüdde, T., Trautwein, C., Groll, J. and Tacke, F. 2012. Peptide-functionalized gold nanorods increase liver injury in hepatitis. *ACS Nano*, 6, 8767–8777.

6. Li, J., Zhu, B., Zhu, Z., Zhang, Y., Yao, X., Tu, S., Liu, R., Jia, S. and Yang, C. J. 2015. Simple and rapid functionalization of gold nanorods with oligonucleotides using an mPEG-SH/Tween 20-assisted approach. *Langmuir*, 31, 7869–7876.

7. Yang, H., Chen, Z., Zhang, L., Yung, W.-Y., Leung, K. C.-F., Chan, H. Y. E. and Choi, C. H. J. 2016. Mechanism for the cellular uptake of targeted gold nanorods of defined aspect ratios. *Small*, 12, 5178–5189.

8. Su, L., Hu, S., Zhang, L., Wang, Z., Gao, W., Yuan, J. and Liu, M. 2017. A fast and efficient replacement of CTAB with MUA on the surface of gold nanorods assisted by a water-immiscible ionic liquid. *Small*, 13, 1602809.

9. del Caño, R., Gisbert-González, J. M., González-Rodríguez, J., Sánchez-Obrero, G., Madueño, R., Blázquez, M. and Pineda, T. 2020. Effective replacement of cetyltrimethylammonium bromide (CTAB) by mercaptoalkanoic acids on gold nanorod (AuNR) surfaces in aqueous solutions. *Nanoscale*, 12, 658–668.

10. MacLeod, M. J., Goodman, A. J., Ye, H.-Z., Nguyen, H. V.-T., Voorhis, T. V. and Johnson, J. A. 2019. Robust gold nanorods stabilized by bidentate N-heterocyclic-carbene–thiolate ligands. *Nature Chemistry*, 11, 57–63.

11. Zhao, Y., Tong, L., Li, Y., Pan, H., Zhang, W., Guan, M., Li, W., Chen, Y., Li, Q., Li, Z., Wang, H., Yu, X-F. and Chu, P. K. 2016. Lactose-functionalized gold nanorods for sensitive and rapid serological diagnosis of cancer. *ACS Applied Materials & Interfaces*, 8, 5813–5820.

12. Nandanan, E., Jana, N. R. and Ying, J. Y. 2008. Functionalization of gold nanospheres and nanorods by chitosan oligosaccharide derivatives. *Advanced Materials*, 20, 2068–2073.

13. Choi, R., Yang, J., Choi, J., Lim, E.-K., Kim, E., Suh, J.-S., Huh, Y.-M. and Haam, S. 2010. Thiolated dextran-coated gold nanorods for photothermal ablation of inflammatory macrophages. *Langmuir*, 26, 17520–17527.

14. García, I., Sanchez-Iglesias, A., Henriksen-Lacey, M., Grzelczak, M., Penadeś, S. and Liz-Marzan, L. M. 2015. Glycans as biofunctional ligands for gold nanorods: Stability and targeting in protein-rich media. *Journal of the American Chemical Society*, 137, 3686–3692.

15. Liu, C. K., Zheng, Y., Lu, X., Thai, T., Lee, N. A., Bach, U. and Gooding, J. J. 2015. Biocompatible gold nanorods: One-step surface functionalization, highly colloidal stability, and low cytotoxicity. *Langmuir*, 31, 4973–4980.

16. Schulz, F., Friedrich, W., Hoppe, K., Vossmeyer, T., Wellera, H. and Lange, H. 2016. Effective PEGylation of gold nanorods. *Nanoscale*, 8, 7296–7308.

17. Bartneck, M., Keul, H. A., Singh, S., Czaja, K., Bornemann, J., Bockstaller, M., Moeller, M., Zwadlo-Klarwasser, G. and Groll, J. 2010. Rapid uptake of gold nanorods by primary human blood phagocytes and immunomodulatory effects of surface chemistry. *ACS Nano*, 4, 3073–3086.

18. Kwon, Y., Choi, Y., Jang, J., Yoon, S. and Choi, J. 2020. NIR laser-responsive PNIPAM and gold nanorod composites for the engineering of thermally reactive drug delivery nanomedicine. *Pharmaceutics*, 12, 204.

19. Kinnear, C., Burnand, D., Clift, M. J. D., Kilbinger, A. F. M, Rothen-Rutishauser, B. and Petri-Fink, A. 2014. Polyvinyl alcohol as a biocompatible alternative for the passivation of gold nanorods. *Angewandte Chemie International Edition*, 53, 12613–12617.

20. Takahashi, H., Niidome, Y., Niidome, T., Kaneko, K., Kawasaki, H. and Yamada, S. 2006. Modification of gold nanorods using phosphatidylcholine to reduce cytotoxicity. *Langmuir*, 22, 2–5.

21. Matthews, J. R., Payne, C. M. and Hafner, J. H. 2015. Analysis of phospholipid bilayers on gold nanorods by plasmon resonance sensing and surface-enhanced Raman scattering. *Langmuir*, 31, 9893–9900.

22. Mehtala, J. G., Zemlyanov, D. Y., Max, J. P., Kadasala, N., Zhao, S. and Wei, A. 2014. Citrate-stabilized gold nanorods. *Langmuir*, 30, 13727–13730.

23. Sreeprasad, T. T. S. and Pradeep, T. 2011. Reversible assembly and disassembly of gold nanorods induced by EDTA and its application in SERS. *Langmuir*, 27, 3381–3390.

24. Wang, J., Dong, B., Chen, B., Jiang, Z. and Song, H. 2012. Selective photothermal therapy for breast cancer with targeting peptide modified gold nanorods. *Dalton Transactions*, 41, 11134–11144.

25. Wang, L., Liu, Y., Li, W., Jiang, X., Ji, Y., Wu, X., Xu, L., Qiu, Y., Zhao, K., Wei, T., Li, Y., Zhao, Y. and Chen, C. 2011. Selective targeting of gold nanorods at the mitochondria of cancer cells: Implications for cancer therapy. *Nano Letters*, 11, 772–780.

26. Kah, J. C. Y., Chen, J., Zubieta, A. and Hamad-Schifferli, K. 2012. Exploiting the protein corona around gold nanorods for loading and triggered release. *ACS Nano*, 6, 6730–6740.

27. Tebbe, M., Kuttner, C., Mannel, M., Fery, A. and Chanana, M. 2015. Colloidally stable and surfactant-free protein-coated gold nanorods in biological media. *ACS Applied Materials & Interfaces*, 7, 5984–5991.

28. Yeo, E. L. L., Cheah, J. U-J., Lim, B. Y., Thong, P. S. P., Soo, K. C. and Kah, J. C. Y. 2017. Protein corona around gold nanorods as a drug carrier for multimodal cancer therapy. *ACS Biomaterials Science & Engineering*, 3, 1039–1050.

29. Kim, D. H., Wei, A. and Won, Y.-Y. 2012. Preparation of super-stable gold nanorods via encapsulation into block copolymer micelles. *ACS Applied Materials & Interfaces*, 4, 1872–1877.

30. Obare, S. O., Jana, N. R. and Murphy, C. J. 2001. Preparation of polystyrene- and silica-coated gold nanorods and their use as templates for the synthesis of hollow nanotubes. *Nano Letters*, 1, 601–603.

31. Pastoriza-Santos, I., Pe´rez-Juste, J. and Liz-Marza´n, L. M. 2006. Silica-coating and hydrophobation of CTAB-stabilized gold nanorods. *Chemistry of Materials*, 18, 10, 2465–2467.

32. Abadeer, N. S., Brennan, M. R., Wilson, W. L. and Murphy, C. J. 2014. Distance and plasmon wavelength dependent fluorescence of molecules bound to silica-coated gold nanorods. *ACS Nano*, 8, 8392–8406.

33. Pellas, V., Blanchard, J., Guibert, C., Krafft, J.-M., Miche, A., Salmain,` M. and Boujday, S. 2021. Gold nanorod coating with silica shells having controlled thickness and oriented porosity: Tailoring the shells for biosensing. *ACS Applied Nano Materials*, 4, 9842–9854.

34. Gole, A. and Murphy, C. J. 2005. Polyelectrolyte-coated gold nanorods: Synthesis, characterization and immobilization. *Chemistry of Materials*, 17, 1325–1330.

35. Alkilany, A. M., Boulos, S. P., Lohse, S. E., Thompson, L. B. and Murphy, C. J. 2014. Homing peptide-conjugated gold nanorods: The effect of amino acid sequence display on nanorod uptake and cellular proliferation. *Bioconjugate Chemistry*, 25, 1162–1171.

36. Basiruddin, S. K., Saha, A., Pradhan, N. and Jana, N. R. 2010. Functionalized gold nanorod solution via reverse micelle based polyacrylate coating. *Langmuir*, 26, 7475–7481.

37. Kaneti, Y. V., Chen, C., Liu, M., Wang, X., Yang, J. L., Taylor, R. A., Jiang, X. and Yu, A. 2015. Carbon-coated gold nanorods: A facile route to biocompatible materials for photothermal applications. *ACS Applied Materials & Interfaces*, 7, 25658–25668.

38. Hinman, J. G., Turner, J. G., Hofmann, D. M. and Murphy, C. J. 2018. Layer-by-layer synthesis of conformal metal–organic framework shells on gold nanorods. *Chemistry of Materials*, 30, 7255–7261.

39. Li, Z., Huang, P., Zhang, X., Lin, J., Yang, S., Liu, B., Gao, F., Xi, P., Ren, Q. and Cui, D. 2009. RGD-conjugated dendrimer-modified gold nanorods for in vivo tumor targeting and photothermal therapy. *Molecular Pharmaceutics*, 7, 94–104.

40. Liao, H. and Hafner, J. H. 2005. Gold nanorod bioconjugates. *Chemistry of Materials*, 17, 4636–4641. Chakraborty, A., Dalal, C. and Jana, N. R. 2018. Colloidal nanobioconjugate with complementary surface chemistry for cellular and subcellular targeting. *Langmuir*, 34, 13461–13471.

Plasmonic Property of Gold Nanorod for Optical Probe

5

5.1 INTRODUCTION

Optical detection of chemicals and biochemicals is the most popular approach in chemical and biological science. In particular this approach has been utilized for monitoring environmental pollutants and medical conditions, observing biological events, and understanding biological processes. Moreover, different types of optical detection approaches can be adapted depending on the requirements, for example, colorimetric detection, spectrophotometric or fluorimetric detection, fluorescence imaging-based detection, etc.

Gold nanorod is a very powerful optical detection probe for four reasons.[1-7] First, it has an intense plasmon absorption band that is tunable from a visible to near-infrared region. This offers various optical-based detection applications in the visible range. Second, gold nanorod has colloidal properties. This makes it easier for different chemical/bio-environment to capture and transform gold nanorods into optical signals. Third, the small size of nanorods means they can penetrate cells and subcellular compartments; and they can then be used to monitor intracellular biochemical activity. Fourth, gold nanorods have high chemical stability allowing their application in different adverse environments, and they can be processed for different detection platforms. Due to these advantages, different optical detection platforms use gold nanorods as probes[1-7] (see Table 5.1). This chapter will discuss various optical detection mentions that use gold nanorods, including colorimetric detection, spectrophotometric detection, surface-enhanced Raman spectroscopy (SERS)-based detection, fluorescence resonance energy transfer (FRET)-based detection, and super-resolution imaging. These approaches have been used for the detection

DOI: 10.1201/9781003245339-5

TABLE 5.1   Gold Nanorod-Based Optical Detection Approaches with Some Selected Examples.

DETECTION METHOD	FUNCTIONALIZATION WITH	DETECTION MECHANISM	DETECTION OF	REFERENCE
Colorimetric	horseradish peroxidase, glucose oxidase	H_2O_2 based degradation of nanorod	blood sugar	8
Colorimetric	thiocyanate	Fenton-like reaction-mediated etching of gold nanorod	Co^{2+} ions	9
Colorimetric	antibody, ascorbic acid 2-phosphate, iodate	iodine-mediated etching of gold nanorod	alkaline phosphatase	10
Colorimetric	beta-galactosidase	enzymatic reaction-induced silver deposition on the surface of gold nanorods	Escherichia coli	11
Colorimetric	DNA crosslinked hydrogel	responsive aggregation of nanorods in hydrogel	cocaine	12
Colorimetric	Hg⁺ or PEG-thiol	gold amalgam and end-to-end assembly	ultratrace Hg^{2+}	13, 15
Colorimetric	lipase	enzymatic reaction-assisted gold deposition on gold nanorod	lipase activity	14
Colorimetric	antibody	color of nanorod in lateral flow immunoassay	protein	16
Plasmonic-based	–	oxidative dissolution	persulfate, cyanide	17
Plasmon-based	cysteine/glutathione at either end	linear assembly	cysteine and glutathione	18
Plasmonic-based	antibody	change of plasmon band	multiple proteins	19
Plasmonic-based	lactose	change of plasmon band after binding with cancer biomarker	serological diagnosis of cancer	20

(Continued)

TABLE 5.1 (CONTINUED)   Gold Nanorod-Based Optical Detection Approaches with Some Selected Examples.

DETECTION METHOD	FUNCTIONALIZATION WITH	DETECTION MECHANISM	DETECTION OF	REFERENCE
Plasmonic-based	–	inhibitory effect of arsenic(III) on iron(II)-mediated oxidative shortening of gold nanorods	arsenic(III)	21
Plasmonic-based	peptide	change of plasmon band	biological event	22
Plasmonic-based (Chip-based)	antibody	change of refractive index and LSPR upon analyte binding	antigen	23
SERS	silver coating	plasmonic hot spot-based electromagnetic enhancement	cyanide, dye	24
SERS	anti-EGFR antibody	nanorod assembly by cancer cell	cancer cell	25, 26
SERS	polymer at either end	nanorod dimers	dye	27
SERS	porous silica shell	reaching of Raman probe to plasmonic core via silica shell	dye	28
SERS	polymer at tip	electromagnetic enhancement	dye	29
SERS	composite with graphene oxide	electromagnetic enhancement	dye	30
SERS	silver coating	electromagnetic enhancement	antibiotic	31
SERS	Raman probe at two ends, antibody at side	electromagnetic enhancement	bacteria	32
SERS	Raman probe	electromagnetic enhancement via nanorod aggregation in microchannel	dye	33

(Continued)

TABLE 5.1 (CONTINUED) Gold Nanorod-Based Optical Detection Approaches with Some Selected Examples.

DETECTION METHOD	FUNCTIONALIZATION WITH	DETECTION MECHANISM	DETECTION OF	REFERENCE
FRET	glutathione-capped fluorescent Au cluster	"Turn on" of quenched emission after analyte binding	glutathione S-transferase	34
Super-resolution imaging	fluorophore labeled double-stranded DNA	steady background luminescence of the gold nanorod as one emission source	heterogeneity in DNA binding to nanorods	35,36

of a wide variety of chemicals/biochemicals including toxic metal ions, amino acids, antibiotics, reactive oxygen species, blood sugar, enzyme, cancer cells, and bacteria.

5.2 COLORIMETRIC AND SPECTROPHOTOMETRIC DETECTION

The strong, intense, and length-dependent plasmon band of gold nanorod in ranges 500–1200 nm has been used for colorimetric or plasmon-based detection applications.[8–23] The detection principle involves either directly using the plasmon band or changing it during the interaction with the analyte (see Table 5.1). The most commonly used detection application requires changing the plasmon band (see Figure 5.1). Three different processes are adapted for changing the plasmon band in detection applications. First, oxidative etching/ dissolution of nanorod is adapted for blue shifting the longitudinal plasmon band. This approach is used for the detection of various oxidizing agents such as metal ions,[9,21] persulfates,[17] reactive oxygen species, or enzymes that generate reactive oxygen species.[8–10] For example, H_2O_2-based chemical etching of gold nanorods induces a decrease in the nanorod aspect ratio and changes in plasmonic properties. Gold nanorod is functionalized with two enzymes such as horseradish peroxidase and glucose oxidase. The glucose oxidase generates H_2O_2 in the presence of glucose, and horseradish peroxidase induces oxidative etching of nanorods in the presence of H_2O_2. This allows optical/colorimetric detection of glucose in micromolar concentration.[8] Similarly, Fenton-like reaction-mediated etching of gold nanorods is used for sensitive visual detection of Co^{2+} ions. Thiocyanate functionalized colloidal gold nanorod is subjected to hydrogen peroxide and Co^{2+} ions where Co^{2+} ions trigger the Fenton-like reaction and generate superoxide radical. As a result, the gold nanorods are gradually etched preferentially along the longitudinal direction. This is accompanied by a color change from green to red that is used for Co^{2+} ion detection in nanomolar concentration[9] (Figure 5.1). In another approach, iodine-mediated etching of gold nanorods is used for colorimetric detection of alkaline phosphatase. The alkaline phosphatase hydrolyzes ascorbic acid 2-phosphate into ascorbic acid. The ascorbic acid then reduces iodate to iodine and etches the nanorod. As a result, the solution color changes from blue to red.[10]

Second, deposition of gold/silver/mercury has been used to change the nanorod plasmon band followed by the detection of a biochemical.[11,13–15] For example, beta-galactosidase functionalized gold nanorod is used for enzyme-induced silver deposition on the nanorod surface for the detection of

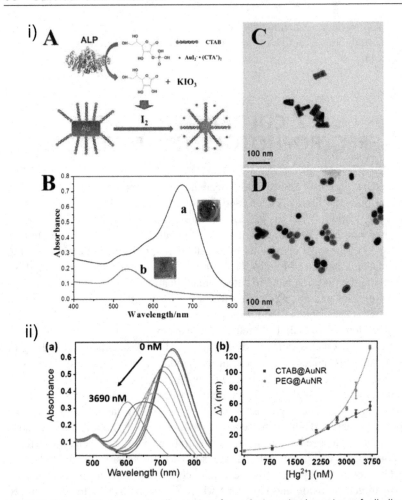

FIGURE 5.1 i) (A) Schematic illustration for colorimetric detection of alkaline phosphatase based on the etching of gold nanorod. (B) Absorption spectra and optical images of gold nanorod before (curve a, blue image) and after (curve b, red image) incubation with alkaline phosphatase. TEM images of the gold nanorod (C) before and (D) after treatment with alkaline phosphatase separately. Reprinted with permission from Zhang, Z. et al. 2015. Iodine-mediated etching of gold nanorods for plasmonic ELISA based on colorimetric detection of alkaline phosphatase. *ACS Applied Materials and Interfaces*, 7, 27639–27645. Copyright 2015 American Chemical Society. ii) (a) Absorption spectra of PEGylated gold nanorod amalgamated at various concentrations of Hg^{2+}. (b) Plots of the mean of longitudinal plasmon shifts of PEGylated and CTAB-capped nanorod as a function of concentration of Hg^{2+}. Reprinted with permission from Crockett, J. R. et al. 2021. Plasmonic detection of mercury via amalgamation on gold nanorods coated with PEG-thiol. *ACS Applied Nano Materials*, 4, 1654–1663. Copyright 2021 American Chemical Society.

Escherichia coli.[11] Similarly, lipase functionalized nanorod is used for enzyme-assisted gold deposition and for the detection of lipase activity.[14] In other work, mercury detection has been achieved at ultralow concentration via the formation of gold amalgam[13,15] (Figure 5.1).

Third, analyte-induced nanorod aggregation and change of plasmon band are used for detection applications.[12,18–20,22] For example, linear assembly of nanorods via the attachment of cysteine/glutathione at either end of nanorods has been used for their detection.[18] In other work, DNA crosslinked hydrogel has been designed for cocaine-responsive aggregation of nanorod in hydrogel and used for colorimetric detection of cocaine.[12] Similarly, antibody/peptide/carbohydrate functionalized nanorods are used for the detection of antigen and biological events via change of the nanorod plasmon band upon the analyte binding event.[19,20,22] Here the plasmon band of gold nanorod changes upon the analyte binding and is used for highly sensitive detection.

5.3 SURFACE-ENHANCED RAMAN SPECTROSCOPY (SERS)-BASED DETECTION

The plasmonic property of gold nanorods has attracted significant attention due to their utilization in surface-enhanced Raman spectroscopy (SERS)-based detection of chemicals and biochemicals.[24–33] SERS requires the attachment of an analyte at the surface of plasmonic nanoparticles. Under laser excitation, the analyte can experience an enhanced electromagnetic field that increases the nanoparticles' vibrational Raman signals. More specifically, the analyte should localize at electromagnetic hot spots that are produced at the junction between Au/Ag nanoparticle dimers, sharp ends of Au/Ag nanostars, either end of Au/Ag nanorods, and selected positions of aggregated Au/Ag nanoparticles. Gold nanorod has a unique advantage as a substrate for SERS as it has a tunable plasmonic property in the range of 500–1200 nm and thus a wide variety of lasers can be used for their plasmonic excitation and SERS study. In addition, either end of a nanorod can be used for plasmonic hot spot-based SERS.

Table 5.1 highlight the SERS-based detection application of gold nanorods. It includes the detection of toxic chemicals, biochemicals, antibiotics, cancer cells, and bacteria. The most common SERS approach involves the adsorption of dye (as the Raman probe) at the nanorod surface followed by aggregation/assembly of nanorods.[24,27,33] Random aggregation of nanorods can be achieved by adding salt or using microchannels.[24,33] Although this approach can easily generate electromagnetic hot spots, there will be a reduction in the number of hot spots. In contrast, end-to-end assembly of nanorods has been achieved via end-capping of polymers and for generating more effective and increased numbers of electromagnetic hot spots.[27,29] In another approach, a Raman probe is attached at either end of the nanorod, and an antibody is attached to the nanorod sides. This can be used for SERS-based detection of bacteria[32] (see Figure 5.2). Alternatively, a silver coating can be used to enhance the SERS signal of gold nanorods[24,31] or to form a composite of graphene-oxide coated gold nanorods.[30] In another approach, EGFR antibody and Raman probe functionalized gold nanorod is used for the detection of cancer cells. In this application, the cancer cells assemble the nanorods and then detected using SERS.[25,26]

5.4 OTHER OPTICAL DETECTION METHODS

Several other optical detection methods have been developed using gold nanorods. These include fluorescence resonance energy transfer (FRET)-based detection and super-resolution imaging.[34–36] In particular, the plasmonic property and emission property of gold nanorods are used for such applications. For example, FRET-based detection of glutathione S-transferase has been developed using gold nanorods and fluorescent gold clusters.[34] Glutathione-capped gold nanoclusters and amine-terminated gold nanorods are designed for turn-on sensing of glutathione S-transferase. The plasmon band of gold nanorod is tuned for maximum spectral overlap and FRET. The attachment of gluta-thione-capped gold nanoclusters to nanorods leads to fluorescence quenching and FRET. Once glutathione S-transferase is added, it replaces the gold clusters which leads to the reappearance of gold cluster fluorescence. Glutathione S-transferase can be detected at concentrations 2–100 nM and the limit of detection is about 1.5 nM.

In another example, gold nanorod has been used in super-resolution imaging applications. In particular, the super-resolution imaging technique is used to investigate the locations of fluorescently labeled double-stranded DNA around gold nanorods.[35,36] In this super-resolution imaging, it is necessary to have two fluorescent probes; gold nanorod is used as one probe. Within each diffraction-limited spot, two different emission sources are used which include the stochastic fluorescence from DNA-bound fluorophore and the

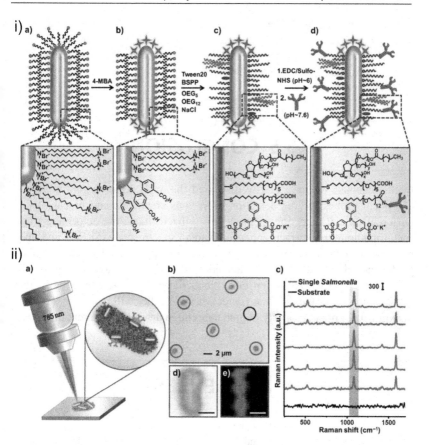

FIGURE 5.2 i) Schematic of steps for the fabrication of gold nanorod-based SERS labels. (a) As-synthesized CTAB-capped gold nanorods, (b) tip-functionalized gold nanorods with the Raman reporter molecule, (c) thiolated PEG modification at side faces of gold nanorods, facilitated by using BSPP, Tween 20, and NaCl, and (d) final gold nanorod labels after conjugation with the bio-targeting moiety (antibody). ii) (a) Schematic representation of SERS analysis of *S. typhimurium* conjugated with the gold nanorod probe under NIR laser illumination (785 nm), (b) optical micrograph of the captured bacteria on the substrate, (c) corresponding SERS spectrum collected from each of the single bacterium spots (circled in gray) and the substrate (circled in black), (d) optical image of the single bacterium, and (e) associated Raman image created based on the intensity at 1078 cm⁻¹ to the background (the scale bar represent 1 μm). Reprinted with permission from Pardehkhorram, R. et al. 2021. Functionalized gold nanorod probes: A sophisticated design of SERS immunoassay for bio-detection in complex media. *Analytical Chemistry*, 93, 12954–12965. Copyright 2021 American Chemical Society.

steady background luminescence of the gold nanorod. Subtracting the average gold nanorod luminescence contribution, the contribution from the fluorescent labels is isolated and provides information on DNA binding heterogeneity across the surface of the nanorod.

5.5 CONCLUSION

In this chapter, we have briefly discussed the application of gold nanorods as optical probes for the detection of various chemicals and biochemicals. Different optical detection methods are utilized, including colorimetry and spectrophotometry, surface-enhanced Raman spectroscopy (SERS), and fluorescence resonance energy transfer (FRET). Appropriately functionalized gold nanorods are used for the detection of toxic metal ions, amino acids, antibiotics, reactive oxygen species, blood sugar, enzyme, cancer cells, and bacteria. It is expected that these detection approaches will inspire researchers to exploit the optical property of gold nanorod in detecting different materials.

REFERENCES

1. Vigderman, L., Khanal, B. P. and Zubarev, E. R. 2012. Functional gold nanorods: Synthesis, self-assembly, and sensing applications. *Advanced Materials*, 24, 4811–4841.
2. Caoa, J., Suna, T. and Grattan, K. T. V. 2014. Gold nanorod-based localized surface plasmon resonance biosensors: A review. *Sensors and Actuators B*, 195, 332–351.
3. Jayabal, S., Pandikumar, A., Lim, H. N., Ramaraj, R., Sund, T. and Huang, N. M. 2015. A gold nanorod-based localized surface plasmon resonance platform for the detection of environmentally toxic metal ions. *Analyst*, 140, 2540.
4. Scarabelli, L., Hamon, C. and Liz-Marzan, L. M. 2017. Design and fabrication of plasmonic nanomaterials based on gold nanorod supercrystals. *Chemistry of Materials*, 29, 15–25.
5. Wei, W., Bai, F. and Fan, H. 2019. Oriented gold nanorod arrays: Self-assembly and optoelectronic applications. *Angewandte Chemie International Edition*, 58, 11956–11966.
6. Murphy, C. J., Chang, H.-H., Falagan-Lotsch, P., Gole, M. T., Hofmann, D. M., Hoang, K. N. L., McClain, S. M., Meyer, S. M., Turner, J. G., Unnikrishnan, M., Wu, M., Zhang, X. and Zhang, Y. 2019. Virus-sized gold nanorods: Plasmonic particles for biology. *Accounts for Chemical Research*, 52, 2124–2135.

7. Gorbunova, M., Apyari, V., Dmitrienko, S. and Zolotov, Y. 2020. Gold nanorods and their nanocomposites: Synthesis and recent applications in analytical chemistry. *Trends in Analytical Chemistry*, 130, 115974.

8. Saa, L., Coronado-Puchau, M., Pavlov, V. and Liz-Marzan, L. M. 2014. Enzymatic etching of gold nanorods by horseradish peroxidase and application to blood glucose detection. *Nanoscale*, 6, 7405–7409.

9. Zhang, Z., Chen, Z., Pan, D. and Chen, L. 2015. Fenton-like reaction-mediated etching of gold nanorods for visual detection of Co^{2+}. *Langmuir*, 31, 643–650.

10. Zhang, Z., Chen, Z., Wang, S., Cheng, F. and Chen, L. 2015. Iodine-mediated etching of gold nanorods for plasmonic ELISA based on colorimetric detection of alkaline phosphatase. *ACS Applied Materials and Interfaces*, 7, 27639–27645.

11. Chen, J., Jackson, A. A., Rotello, V. M. and Nugen, S. R. 2016. Colorimetric detection of escherichia coli based on the enzyme-induced metallization of gold nanorods. *Small*, 12, 2469–2475.

12. Mao, Y., Li, J., Yan, J., Ma, Y., Song, Y., Tian, T., Liu, X., Zhu, Z., Zhou, L. and Yang, C. 2017. A portable visual detection method based on a target-responsive DNA hydrogel and color change of gold nanorods. *Chemical Communications*, 53, 6375–6378.

13. Chen, L., Lu, L., Wang, S. and Xia, Y. 2017. Valence states modulation strategy for picomole level assay of Hg^{2+} in drinking and environmental water by directional self-assembly of gold nanorods. *ACS Sensors*, 2, 781–788.

14. Zhang, H., Wu, S., Zhang, L., Jiang, L., Huo, F. and Tian, D. 2019. High-resolution colorimetric detection of lipase activity based on enzyme-controlled reshaping of gold nanorods. *Analytical Methods*, 11, 2286.

15. Crockett, J. R., Win-Piazza, H., Doebler, J. E., Luan, T. and Bao, Y. 2021. Plasmonic detection of mercury via amalgamation on gold nanorods coated with PEG-thiol. *ACS Applied Nano Materials*, 4, 1654–1663.

16. Pang, R., Zhu, Q., Wei, J., Wang, Y., Xu, F., Meng X. and Wang, Z. 2021. Development of a gold-nanorod-based lateral flow immunoassay for a fast and dual-modal detection of C-reactive protein in clinical plasma samples. *RSC Advances*, 11, 28388.

17. Jana, N. R., Gearheart, L., Obare, S. O. and Murphy, C. J. 2002. Anisotropic chemical reactivity of gold spheroids and nanorods, *Langmuir*, 18, 3, 922–927.

18. Sudeep, P. K., Joseph, S. T. S and Thomas, K. G. 2005. Selective detection of cysteine and glutathione using gold nanorods. *Journal of the American Chemical Society*, 127, 6516–6517.

19. Yu, C. and Irudayara, J. 2007. Multiplex biosensor using gold nanorods. *Analytical Chemistry*, 79, 572–579.

20. Zhao, Y., Tong, L., Li, Y., Pan, H., Zhang, W., Guan, M., Li, W., Chen, Y., Li, Q., Li, Z., Wang, H., Yu, X.-F. and Chu, P. K. 2016. Lactose-functionalized gold nanorods for sensitive and rapid serological diagnosis of cancer. *ACS Applied Materials and Interfaces*, 8, 5813–5820.

21. Das, A., Mohanty, S. and Kuanr, B. K. 2019. Label-free gold nanorod-based plasmonic sensing of arsenic(III) in contaminated water. *Analyst*, 144, 4708–4718.

22. Beiderman, M., Ashkenazy, A., Segal, E., Barnoy, E. A., Motiei, M., Sadan, T., Salomon, A., Rahimipour, A., Fixler, D. and Popovtzer, R. 2020. Gold nanorod-based bio-barcode sensor array for enzymatic detection in biomedical applications. *ACS Applied Nano Materials*, 3, 8414–8423.

23. Mayer, K. M., Lee, S., Liao, H., Rostro, B. C., Fuentes, A., Scully, P. T., Nehl, C. L. and Hafner, J. H. 2008. A label-free immunoassay based upon localized surface plasmon resonance of gold nanorods. *ACS Nano*, 2, 687–692.

24. Jana, N. R. and Pal, T. 2007. Anisotropic metal nanoparticles for use as surface-enhanced Raman substrates. *Advanced Materials*, 19, 1761–1765.

25. Huang, X., El-Sayed, I. H., Qian, W. and El-Sayed, M. A. 2007. Cancer cells assemble and align gold nanorods conjugated to antibodies to produce highly enhanced, sharp, and polarized surface Raman spectra: A potential cancer diagnostic marker. *Nano Letters*, 7, 1591–1597.

26. Jokerst, J. V., Cole, A. J., Van de Sompel, D. and Gambhir, S. S. 2012. Gold nanorods for ovarian cancer detection with photoacoustic imaging and resection guidance via Raman imaging in living mice. *ACS Nano*, 6, 10366–10377.

27. Stewart, A. F., Lee, A., Ahmed, A., Ip, S., Kumacheva, E. and Walker, G. C. 2014. Rational design for the controlled aggregation of gold nanorods via phospholipid encapsulation for enhanced Raman scattering. *ACS Nano*, 8. 5462–5467.

28. Kang, H. and Haynes, C. 2019. Interactions between silica-coated gold nanorod substrates and hydrophobic analytes in colloidal surface-enhanced Raman spectroscopy. *The Journal of Physical Chemistry C*, 123, 24685–24697.

29. Yilmaz, H., Bae, S. H., Cao, S., Wang, Z., Raman, B. and Singamaneni, S. 2019. Gold-nanorod-based plasmonic nose for analysis of chemical mixtures. *ACS Applied Nano Materials*, 2, 3897–3905.

30. Ponlamuangdee, K., Hornyak, G. L., Borab, T. and Bamrungsap, S. 2020. Graphene oxide/gold nanorod plasmonic paper: A simple and cost-effective SERS substrate for anticancer drug analysis. *New Journal of Chemistry*, 44, 14087–14094.

31. Peng, X., Li, D., Li, Y., Xing, H. and Deng, W. 2021. Plasmonic tunable Ag-coated gold nanorod arrays as reusable SERS substrates for multiplexed antibiotics detection. *Journal of Materials Chemistry B*, 9, 1123–1130.

32. Pardehkhorram, R., Alshawawreh, F. A., Goncales, V. R., Lee, N. A., Tilley, R. D. and Gooding, J. J. 2021. Functionalized gold nanorod probes: A sophisticated design of SERS immunoassay for biodetection in complex media. *Analytical Chemistry*, 93, 12954–12965.

33. Bär, J., de Barros, A., de Camargo, D. H. S., Pereira, M. P., Merces, L., Shimizu, F. M., Sigoli, F. A., Bufon, C. C. B. and Mazali, I. O. 2021. Silicon microchannel-driven Raman scattering enhancement to improve gold nanorod functions as a SERS substrate toward single-molecule detection. *ACS Applied Materials and Interfaces*, 13, 36482–36491.

34. Qin, L., He, X., Chen, L. and Zhang, Y. 2015. Turn-on fluorescent sensing of gluta-thione Stransferase at near infrared region based on FRET between gold nanoclusters and gold nanorods. *ACS Applied Materials and Interfaces*, 7, 5965–5971.

35. Blythe, K. L., Titus, E. J. and Willets, K. A. 2015. Comparing the accuracy of reconstructed image size in super-resolution imaging of fluorophore-labeled gold nanorods using different fit models. *Journal of the Physical Chemistry C*, 119, 19333–19343.

36. Blythe, K. L., Titus, E. J. and Willets, K. A. 2015. Effects of tuning fluorophore density, identity, and spacing on reconstructed images in super-resolution imaging of fluorophore labeled gold nanorods. *The Journal of Physical Chemistry C*, 119, 28099–28110.

Gold Nanorod as Imaging Contrast Agent

6

6.1 INTRODUCTION

The unique optical property of gold nanorods has been utilized as contrast agents in various types of imaging.[1,2] These applications include using gold nanorods as a contrast agent in dark-field imaging, two-photon fluorescence imaging, surface-enhanced Raman imaging, photoacoustic imaging, and optical coherence tomography imaging. Gold nanorod acts as a powerful contrast agent or optical probe in those imaging applications for four reasons.[1,2] First, gold nanorod has an intense and tunable plasmon absorption band from a visible to near-infrared (NIR) region. This property offers strongly enhanced absorption and scattering of light that are used in imaging contrast. Second, increased anisotropy of gold nanorods enhances two-photon luminescence cross sections. This means gold nanorods can be used in two-photon imaging application. Third, colloidal and functional gold nanorod can be designed for labeling of different biological systems. This offers options to enhance the optical signals of specific biocomponents suitable for imaging applications. Fourth, the small size of nanorods means they can penetrate cells and subcellular compartments which allows the imaging of intracellular components.

This chapter will discuss the utilization of gold nanorods in different imaging modalities along with their biomedical applications. These approaches have been used for imaging-based detection in a wide variety of biosystems.

6.2 GOLD NANOROD AS DARK-FIELD CONTRAST AGENT

In dark-field microscopy, a sample is illuminated with white light at an angle and scattered light from the sample is collected under a dark background.[3]

DOI: 10.1201/9781003245339-6

This generates a bright image against a dark background. Dark-field microscopy is ideally used to illuminate unstained samples that appear bright against a dark background. This microscopy uses a special condenser that directs a cone of light away from the objective lens so that only scattered light will be collected without any background light. The advantage of dark-field microscopy is that it requires a simple setup and the quality of images obtained from this technique is impressive. However, the limitation of dark-field microscopy is the low diffracted light seen in the final image and the sample must be strongly illuminated. In addition, interpretation of dark-field images must be done with great care, particularly for biological specimen. In this respect, the use of gold nanorod as a contrast agent in dark-field microscopy can significantly enhance image quality. This is due to the strong light-scattering property of gold nanorod. Thus colloidal gold nanorod-based contrast agents have been designed that can label biological components and then be used for dark-field imaging. Here we will describe some of the selected examples.

Due to strong electric fields at the surface, the absorption and scattering of electromagnetic radiation by gold nanorods are strongly enhanced. These unique properties provide the potential of designing novel optically active reagents for simultaneous molecular imaging and photothermal cancer therapy.[4] Gold nanorods can absorb and scatter in the NIR region and it has been used as novel contrast agents for both molecular imaging and photothermal cancer therapy. Nanorod is conjugated with anti-epidermal growth factor receptor (anti-EGFR) antibodies and selectively labels malignant epithelial cell lines due to the over-expressed EGFR on the cytoplasmic membrane of the malignant cells. Next, strongly scattered red light from gold nanorods in a dark-field is used to visualize malignant cells against nonmalignant cells. After exposure to continuous red laser at 800 nm, malignant cells are photothermally destroyed without affecting nonmalignant cells.

A tiny change in the shape of gold nanorod leads to optical changes in colloidal gold nanorod. This has been used to monitor chemical reactions.[5] The chemical reshaping of gold nanorod induced by the coupling reaction between Au, ferric chloride, and thiourea is monitored via this approach. In the coupling reaction, Fe^{3+} oxidizes the nanorod to yield Au(I), which complexes with thiourea and promotes the chemical reshaping of nanorod. This coupling reaction process has been monitored using a light-scattering dark-field microscopy imaging technique. The light scattering undergoes a color change from bright red to yellow and finally to green, and the gold nanorod undergoes a morphological change from rod-shaped to fusiform and finally to spherical.

Hybrid nanoparticles can be used in multimodal imaging/detection and magnetic separation and are considered a useful probe in understanding cellular function. Gold-nanorod-based plasmonic-fluorescent, plasmonic-magnetic, and plasmonic-fluorescent-magnetic hybrid cellular probes have been synthesized.[6] In the hybrid probes, the nanorod component acts as dark-field contrast agent, the quantum dot acts as a fluorescence probe, and the magnetic iron oxide offers magnetic separation. Oleyl- and glucose-functionalized hybrid nanoprobes have been synthesized and used as dual imaging probes and for magnetic separation. In another approach, folate-functionalized gold nanorod has been synthesized and used as a dark-field imaging probe for the detection of cancer cells and tissues.[7]

6.3 GOLD NANOROD AS TWO-PHOTON FLUORESCENCE IMAGING PROBE

Two-photon fluorescence microscopy is a fluorescence imaging technique that allows visualization of living tissue up to about one millimeter in thickness, with micron-scale resolution.[8] It involves simultaneous excitation by two photons that have longer wavelengths than the emitted light. Two-photon fluorescence microscopy typically uses near-infrared excitation light. The advantage of this imaging approach is that the background signal is strongly suppressed and the use of infrared light increases the penetration depth.

Gold nanorod has been used as a bright contrast agent for two-photon luminescence imaging of cancer cells in a three-dimensional tissue phantom down to 75 μm deep.[9] The two-photon luminescence intensity from gold-nanorod-labeled cancer cells is 3 orders of magnitude brighter than the two-photon autofluorescence emission intensity at 760 nm excitation light. The strong signal, resistance to photobleaching, and biocompatibility make gold nanorods an attractive contrast agent for two-photon imaging of epithelial cancer.

Gold nanorods coated with cetyltrimethylammonium bromide are rapidly internalized by human epithelial carcinoma (KB) cells via a nonspecific uptake mechanism. Internalized nanorods are monitored by two-photon luminescence microscopy and observed to migrate toward the nucleus with a quadratic rate of diffusion.[10] The internalized nanorods are not excreted but form permanent aggregates within cells, which remain healthy and grow to confluence over a 5-day period. Nonspecific nanorod uptake could be greatly reduced by displacing the cetyltrimethylammonium bromide surfactant layer with poly(ethylene glycol).

In another work, folate-targeted gold nanorods are proposed as selective theranostic agents for osteosarcoma treatment.[11] An amphiphilic polymer functionalized with folic acid was used as a coating agent for nanorods.

The obtained polymer-coated nanorod is loaded with anticancer drug nutlin-3 and is able to deliver the drug efficiently in different physiological media. The ability of the proposed systems to selectively kill tumor cells was tested. The work also demonstrated how nanosystems efficiently control drug release upon NIR laser irradiation and how they act as excellent hyperthermia agents and two-photon luminescence imaging contrast agents.

6.4 GOLD NANOROD AS PHOTOACOUSTIC IMAGING PROBE

Photoacoustic imaging or optoacoustic imaging is a biomedical imaging approach that utilizes the photoacoustic effect of gold nanorods.[12] In this approach a sample is exposed with non-ionizing laser pulses and part of the energy is absorbed and converted into heat. This leads to transient thermoelastic expansion and is associated with ultrasonic emission. The generated ultrasonic wave is used to produce an image.

The intrinsic optical absorption contrast and diffraction-limited high spatial resolution of ultrasound make this a promising imaging approach for a wide variety of biomedical applications. Photoacoustic microscopy has important applications in functional imaging. This imaging approach can detect changes in oxygenated/deoxygenated hemoglobin in small vessels and soft tissues in the brain with different optical absorption properties.

Improved imaging modalities are critically needed for optimizing stem cell therapy. Techniques with real-time content to guide and quantitate cell implantation are especially important in applications such as musculoskeletal regenerative medicine. Silica-coated gold nanorods have been used as contrast agents for photoacoustic imaging and quantitation of mesenchymal stem cells in rodent muscle tissue.[13] The silica coating increased the uptake of gold into the cell more than 5-fold without cytotoxicity or proliferation. The low background of the technique allowed imaging down to 100,000 cells in vivo. The spatial resolution is 340 μm, and the temporal resolution is 0.2 s, which is at least an order of magnitude below existing cell imaging approaches. This approach has significant advantage over traditional cell imaging techniques like positron emission tomography and magnetic resonance imaging, including real-time monitoring of stem cell therapy.

Improved imaging approaches are needed for ovarian cancer screening, diagnosis, staging, and resection guidance. A combined photoacoustic/Raman approach using gold nanorods was proposed as a molecular imaging agent. Gold nanorod with an aspect ratio of 3.5 were selected for their highest *ex vivo* and *in vivo* photoacoustic signals and used to image subcutaneous xenografts of ovarian cancer cell lines in living mice.[14] The maximum photoacoustic signal was observed within 3 h and the increased signal persisted for at least two days postadministration. There was a linear relationship between the photoacoustic signal and the concentration of injected molecular imaging agent with a calculated limit of detection of 0.40 nM nanorod in the cell line. The same molecular imaging agent could be used for clear visualization of the margin between tumor and normal tissue and tumor debulking via surface-enhanced Raman spectroscopy (SERS) imaging.

An optoacoustic imaging agent was designed by self-assembling gold nanorods onto DNA-origami nanostructures.[15] The gold-nanorod–DNA-nanostructure hybrid, which combines the advantages of gold nanorods with the DNA-origami structure, serves as a unique probe and an efficient contrast agent in optoacoustic imaging and allows improved imaging quality and decreased dose. Moreover, nanorods responded to NIR irradiation for photothermal therapy and effectively inhibited tumor regrowth and prolonged the survival of diseased mice. Since the optoacoustic imaging modality utilizes non-ionizing radiation, it is therefore safe for patients and medical staff. The combination of optoacoustic imaging with a theranostic agent will encourage further studies investigating the potential of gold nanorod in clinical applications in the near future.

In another example, multifunctional nanoparticles were designed that had high gene transfection activity, low cytotoxicity, photoacoustic imaging abilities, and photothermal therapeutic properties.[16] This involves conjugating low-molecular-weight polyethylenimine onto the surfaces of gold nanorods. Results revealed that the gene transfection efficiency of the prepared polyethylenimine-modified gold nanorods is higher. Material can be a potential photoacoustic and photothermal reagent to evaluate the pharmacokinetics, biodistributions, and antitumor effects of gene/drug nanoparticles.

The visualization of choroidal neovascularization is a major challenge to improving treatment outcomes for patients with age-related macular degeneration. The limitations of current imaging techniques include limited penetration depth, spatial resolution, and sensitivity and difficulty in visualizing components from the healthy microvasculature. RGD-conjugated gold nanorod-based multimodal photoacoustic microscopy (PAM) and optical coherence tomography are used to distinguish the margin of choroidal neovascularization in living rabbits[17] (see Figure 6.1). Intravenous administration of nanorod into rabbit models resulted in signal enhancements of 27-fold in photoacoustic microscopy and 170% in optical coherence tomography. This molecular imaging technique is a promising tool for the precise imaging of

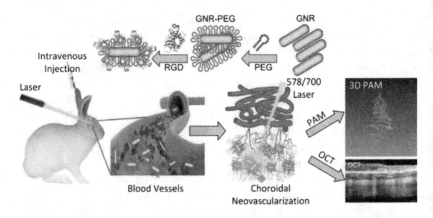

FIGURE 6.1 Schematic of experimental process of RGD-conjugated gold nanorod-based multimodal photoacoustic microscopy (PAM) and optical coherence tomography (OCT) that are used to distinguish the margin of choroidal neovascularization in living rabbits. Reprinted with permission from Nguyen, et al. 2021. Gold nanorod enhanced photoacoustic microscopy and optical coherence tomography of choroidal neovascularization. *ACS Applied Materials and Interfaces*, 13, 40214–40228. Copyright 2021 American Chemical Society.

choroidal neovascularization as well as the evaluation of the pathophysiology *in vivo* without destruction of tissue.

Photoacoustic imaging holds great promise as a non-invasive imaging modality. Gold nanorods with absorption in the second near-infrared (NIR-II) window have emerged as excellent photoacoustic probes because of their tunable optical absorption, surface modifiability, and low toxicity. Gold nanorod–melanin nanohybrids with a tunable polydopamine coating showed a three-fold higher photoacoustic signal due to the increased optical absorption, cross-sectional area, and thermal confinement.[18] This strategy affords a rational design for robust photoacoustic imaging probes and provides more opportunities for other types of photomediated biomedicines, such as photothermal and photodynamic regimens.

6.5 GOLD NANOROD AS SURFACE-ENHANCED RAMAN IMAGING PROBE

Surface-enhanced Raman spectroscopy (SERS) utilizes the plasmonic properties of gold and silver nanoparticles to enhance the Raman signal.[19] In this approach, Raman probes situated at the surface of plasmonic nanoparticles experience higher electromagnetic fields under laser excitation. This leads to the enhancement of the vibrational Raman signal by $>10^6$ fold. A wide variety of SERS probes have been designed that use gold/silver nanoparticles of different sizes/shapes. Gold nanorod is a unique SERS probe as it has tunable plasmon bands that can be excited with different lasers. The SERS-based detection application of gold nanorods was discussed in Chapter 5. However, in some cases, a SERS signal is used for Raman imaging, which is briefly discussed here. For example, molecule-coded gold nanorod was designed as a platform for multiplexed detection, via an intense surface-enhanced Raman spectroscopy (SERS) effect, and remote-controlled therapy, via photothermal heating.[20] Surface-enhanced Raman-active molecules confer intense Raman signatures that are detectable down to attomolar nanorod concentrations. This material generates photothermal heat that ablates tumors *in vivo*. In addition is can be used for *in vivo* SERS imaging. It is proposed that this approach can be used for simultaneous imaging of >10 spectrally unique SERS-coded nanorod under a single NIR excitation source, providing a novel route toward highly parallel *in vivo* screening of nanoparticle behavior.

6.6 GOLD NANOROD AS IMAGING PROBE FOR OPTICAL COHERENCE TOMOGRAPHY

Optical coherence tomography uses low coherence light for imaging applications with micrometer resolution.[21] It is used for two- and three-dimensional medical imaging. The use of longer wavelengths of light allows deep penetration into the scattering medium. This imaging approach is important in the medical community because it provides information on tissue morphology at less than 10–20 μm resolution which is better than MRI or ultrasound imaging.

Optical coherence tomography can be utilized with significant speckle reduction techniques using highly scattering contrast agents for non-invasive, contrast-enhanced imaging of living tissues at the cellular scale. The advantages of reduced speckle noise and improved targeted contrast can be harnessed to track objects as small as 2 μm *in vivo*, which enables applications for cell tracking and quantification in living subjects. Large gold nanorods have demonstrated use as contrast agents for detecting individual micron-sized polystyrene beads and single myeloma cells in blood circulation using speckle-modulating optical coherence tomography[22] (see Figure 6.2). This approach has been used to detect individual cells within blood *in vivo*. This imaging

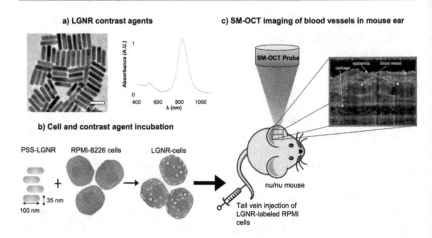

FIGURE 6.2 Schematic of experimental process of optical coherence tomography (OCT) imaging using gold nanorod. (a) Typical TEM and Vis-NIR absorbance spectrum of long gold nanorod (LGNR) used for labeling experiments. (b) LGNRs bio-functionalized with a coating of polystyrene sulfonate (PSS) are incubated with RPMI-8226 cells, resulting in LGNR-labeled cells. (c) Incremental injections of labeled cells are administered into the tail vein of the mouse. Speckle-modulating (SM-OCT) scans at the cross section of blood vessels found in the ear of the mouse, and multiple scans are acquired at the same location to determine the number of cells flowing through the blood vessel over time. Reprinted with permission from Dutta, R. et al. 2019. Real-time detection of circulating tumor cells in living animals using functionalized large gold nanorods. *Nano Letters*, 19, 2334–2342. Copyright 2019 American Chemical Society.

approach has the capability for dynamic detection and quantification of tumor cells circulating in living subjects.

6.7 CONCLUSION

In summary, gold nanorod has been exploited as a contrast agent in various types of imaging such as dark-field imaging, two-photon fluorescence imaging, photoacoustic imaging, surface-enhanced Raman imaging, and optical coherence tomography imaging. These applications utilize the unique optical properties of gold nanorod along with their colloidal properties. In particular, gold nanorods with strong and selective absorption/scattering and enhanced luminescence properties due to shape anisotropy are utilized in all the imaging applications. We expect that more and more imaging application will be explored in future using the unique properties of gold nanorod.

REFERENCES

1. Zheng, J., Cheng, X., Zhang, H., Bai, X., Ai, R., Shao, L. and Wang, J. 2021. Gold nanorods: The most versatile plasmonic nanoparticles. *Chemical Reviews*, 121, 13342–13453.
2. Huang, X., Neretina, S. and El-Sayed, M. A. 2009. Gold nanorods: From synthesis and properties to biological and biomedical applications. *Advanced Materials*, 21, 4880–4910.
3. Gao, P. F., Lei, G. and Huang, C. Z. 2021. Dark-field microscopy: Recent advances in accurate analysis and emerging applications. *Analytical Chemistry*, 93, 4707–4726.
4. Huang, X., El-Sayed, I. H., Qian, W. and El-Sayed, M. A. 2006. Cancer cell imaging and photothermal therapy in the near-infrared region by using gold nanorods. *Journal of the American Chemical Society*, 128, 2115–2120.
5. Zhang, H. Z., Li, R. S., Gao, P. F., Wang, N., Lei, G., Huang, C. Z. and Wang, J. 2017. Real-time dark-field light scattering imaging to monitor the coupling reaction with gold nanorods as an optical probe. *Nanoscale*, 9, 3568–3575.
6. Basiruddin, S. K., Maity, A. R., Saha, A. and Jana, N. R. 2011. Gold-nanorod-based hybrid cellular probe with multifunctional properties. *The Journal of Physical Chemistry C*, 115, 19612–19620.
7. Maity, A. R., Saha, A., Roy, A. and Jana, N. R. 2013. Folic acid functionalized nanoprobes for fluorescence-, dark-field-, and dual-imaging-based selective detection of cancer cells and tissue. *ChemPlusChem*, 78, 259–267.
8. Zong, W., Wu, R., Li, M., Hu, Y., Li, Y., Li, J., Rong, H., Wu, H., Xu, Y., Lu, Y., Jia, H., Fan, M., Zhou, Z., Zhang, Y., Wang, A., Chen, L. and Cheng, H. 2017. Fast high-resolution miniature two-photon microscopy for brain imaging in freely behaving mice. *Nature Methods*, 14, 713–719.
9. Durr, N. J., Larson, T., Smith, D. K., Korgel, B. A., Sokolov, K. and Ben-Yakar, A. 2007. Two-photon luminescence imaging of cancer cells using molecularly targeted gold nanorods. *Nano Letters*, 7, 941–945.
10. Huff, T. B., Hansen, M. N., Zhao, Y., Cheng, J-X. and Wei, A. 2007. Controlling the cellular uptake of gold nanorods. *Langmuir*, 23, 1596–1599.
11. Volsi, A. L., Scialabba, C., Vetri, V., Cavallaro, G., Licciardi, M. and Giammona, G. 2017. Near-infrared light responsive folate targeted gold nanorods for combined photothermal-chemotherapy of osteosarcoma. *ACS Applied Materials and Interfaces*, 9, 14453–14469.
12. Binte, A., Attiaa, E., Balasundarama, G., Moothancherya, M., Dinisha, U. S., Bia, R., Ntziachristosb, V. and Olivo, M. 2019. A review of clinical photoacoustic imaging: Current and future trends. *Photoacoustics*, 16, 100144.
13. Jokerst, J. V., Thangaraj, M., Kempen, P. J., Sinclair, R. and Gambhir, S. S. 2012. Photoacoustic imaging of mesenchymal stem cells in living mice via silica-coated gold nanorods. *ACS Nano*, 6, 5920–5930.

14. Jokerst, J. V., Cole, A. J., Van de Sompel, D. and Gambhir, S. S. 2012. Gold nanorods for ovarian cancer detection with photoacoustic imaging and resection guidance via Raman imaging in living mice. *ACS Nano*, 6, 10366–10377.

15. Du, Y., Jiang, Q., Beziere, N., Song, L., Zhang, Q., Peng, D., Chi, C., Yang, X., Guo, H., Diot, G., Ntziachristos, V., Ding, B. and Tian, J. 2016. DNA-nanostructure–gold-nanorod hybrids for enhanced in vivo optoacoustic imaging and photothermal therapy. *Advanced Materials*, 28, 10000–10007.

16. Chen, J., Liang, H., Lin, L., Guo, Z., Sun, P., Chen, M., Tian, H., Deng, M. and Chen, X. 2016. Gold-nanorods-based gene carriers with the capability of photoacoustic imaging and photothermal therapy. *ACS Applied Materials and Interfaces*, 8, 31558–31566.

17. Nguyen, V.-P., Li, Y., Henry, J., Zhang, W., Wang, X. and Paulus, Y. M. 2021. Gold nanorod enhanced photoacoustic microscopy and optical coherence tomography of choroidal neovascularization. *ACS Applied Materials and Interfaces*, 13, 40214–40228.

18. Yim, W., Zhou, J., Mantri, Y., Creyer, M. N., Moore, C. A. and Jokerst, J. V. 2021. Gold nanorod–melanin hybrids for enhanced and prolonged photoacoustic imaging in the near-infrared-II window. *ACS Applied Materials and Interfaces*, 13, 14974–14984.

19. Zong, C., Xu, M., Xu, L.-J., Wei, T., Ma, X., Zheng, X.-S., Hu, R. and Ren, B. 2018. Surface-enhanced Raman spectroscopy for bioanalysis: Reliability and challenges. *Chemical Reviews*, 118, 4946–4980.

20. von Maltzahn, G., Centrone, A., Park, J.-H., Ramanathan, R., Sailor, M. J., Hatton, T. A. and Bhatia, S. N. 2009. SERS-coded gold nanorods as a multifunctional platform for densely multiplexed near-infrared imaging and photothermal heating. *Advanced Materials*, 21, 3175–3180.

21. Everett, M., Magazzeni, S., Schmoll, T. and Kempe, M. 2021. Optical coherence tomography: From technology to applications in ophthalmology. *Translational Biophotonics*, 3, e202000012

22. Dutta, R., Liba, O., SoRelle, E. D., Winetraub, Y., Ramani, V. C., Jeffrey, S. S., Sledge, G. W. and de la Zerda, A. 2019. Real-time detection of circulating tumor cells in living animals using functionalized large gold nanorods. *Nano Letters*, 19, 2334–2342.

Application of Gold Nanorod in Photothermal Therapy

7

7.1 INTRODUCTION

The application of heat for cancer treatment is widely used since the early 1900s.[1] In this approach, a body containing a tumor is subjected to abnormally high temperature (~40–45°C) compared to its normal physiological temperature (37°C). This approach is commonly known as hyperthermia. Heating can be performed externally using instruments that produce microwaves, radio waves, or ultrasound. In addition, hyperthermia has been used in combination with chemotherapy for more effective cancer therapy.[1]

Gold nanoparticles are suitable for the thermal destruction of cancer due to their photothermal heating ability.[2] The heat generated from gold nanoparticles under light exposure can be utilized to damage/destroy cancerous cells and tissues. While in traditional hyperthermia the heating occurs in all the exposed areas, photothermal heating occurs only around the gold nanoparticles. Thus photothermal heating can be used for targeted therapy without disturbing healthy tissue.

An early example of gold nanoparticle-mediated photothermal cancer therapy used polyethylene glycol (PEG)-terminated gold nanoshell[3] and anti-epidermal growth factor receptor (EGFR) functionalized gold nanospheres.[4] The gold nanoshell consisted of 110 nm silica cores surrounded by 10 nm gold shells with plasmon maxima at 820 nm. Thus near-infrared (NIR) light was used for plasmonic excitation-based photothermal therapy.[3] EGFR-functionalized 40 nm gold nanoparticles have been used to target cancer cells using 514 nm laser irradiation-based cell therapy.[4]

Those early investigations have prompted the growth of gold nanoparticle-based photothermal therapies. Gold nanoparticle-based photothermal cancer

DOI: 10.1201/9781003245339-7

therapy has three specific advantages.[2] First, a wide variety of gold nanoparticles with different sizes and shapes are prepared with varied plasmonic properties. Then they can be exploited for photothermal therapy. Second, gold is chemically inert and biocompatible. Thus gold nanorods can be safely used at clinically relevant concentrations. Third, the size of gold nanoparticles can be tuned in the range of 10–400 nm with appropriate surface chemistry so that they can be used for cell or tumor targeting. However, size, shape, surface chemistry modification of gold nanoparticles, targeting efficiency, and photothermal conditions are extremely critical for better performance.

7.2 PHOTOTHERMAL PROPERTY OF GOLD NANOROD

Gold nanoparticle-based photothermal therapy involves light absorption and subsequent nonradiative energy dissipation in the form of heat.[5] It has been shown that the photothermal heating process in gold nanoparticles excites free electrons in the plasmon band and creates hot electrons. These hot electrons cool through electron–phonon interactions and heat is then transferred from the nanoparticles to their surroundings through phonon–phonon interactions.[5] This results in the surrounding medium increasing in temperature by tens of degrees.

As the photothermal process involves light absorption, the plasmonic property of nanoparticles significantly influence this process. In particular, the position of plasmonic absorption, intensity of plasmon band, and their efficiency in converting light into heat are critical factors.[2] In this respect gold nanorod is unique because their plasmon band can be tuned from a completely visible to NIR range.

The change in temperature from a colloidal gold nanorod with plasmon absorbance maximum at 800 nm has been investigated under irradiation with 800 nm light.[6] Samples were irradiated continuously and temperature was monitored at different positions using thermocouples. Results show that the presence of nanorod increases the temperature up to 45°C and steady-state temperature was reached within 20 min of irradiation.[6] It has been observed that increased concentration of gold nanorods results in a greater temperature increase due to stronger light absorption. However, when used in solution, this process has limitations because increased concentrations of gold nanorods can decrease light penetration, resulting in restricted depth to photothermal heating.[7] In addition increased laser power and irradiation time also increase the heating effect.[6–9]

As relative contributions of absorption vs scattering phenomena in gold nanorods are pivotal to optimizing the efficiency of light-to-heat conversion, a systematic analysis of photothermal properties under low-intensity femtosecond illumination has been conducted. It has been observed that a larger volume of longer gold nanorods in a colloidal solution has comparable performance to much smaller rods in overall photothermal conversion efficiency at identical optical density.[10] However, we need to remember that a smaller nanorod size is better for cellular and subcellular targeting.

7.3 PHOTOTHERMAL EFFECT OF GOLD NANOROD ON LIVING CELLS

What could be the effect of gold nanorod-based photothermal heating on cells? The effects of photothermal heating on cells are similar to traditional hyperthermia. It has been observed that cancer cell death using traditional hyperthermia occurs due to damage to the cell membrane, denaturation of intracellular proteins, influence on DNA replication, and induction of cell death via apoptosis.[1]

Similarly, gold nanoparticle-based photothermal heating leads to cell death via apoptosis and/or necrosis.[11] Apoptosis is a programmed cell death associated with membrane blebbing and nuclear fragmentation. The apoptotic cell components are removed by phagocytes. In contrast, necrosis results from toxicity or damage and is associated with membrane collapse, cell swelling, and cell rupture.

A mechanistic investigation of photothermal injury of KB cells by gold nanorods that are localized at the perinuclear region was conducted.[12] It has been shown that internalized gold nanorods require higher laser intensity as compared to cell surface-bound nanorods and the surface-bound gold nanorods are more effective photothermal transducers. Cell death has been associated with membrane cavitation and disruption of actin filaments with resultant membrane blebbing.[12] The blebbing results in the disruption of the connections between the cell membrane and cytoskeleton followed by cell death via apoptosis.

In another work, real-time monitoring of cancer cell morphology was investigated during gold-nanorod based photothermal heating.[13] Breast cancer cells were incubated with gold nanorods overnight to increase uptake into the endosomes or lysosomes. Generally, a few hundred to a few thousand nanorods enter per cell. Irradiation with a femtosecond pulsed laser led to an explosion of lysosomes. This results in the formation of micron-sized cavities that caused the plasma membrane to rupture. These results suggest that gold nanorods can be used for the photothermal destruction of cells. However, cell targeting, subcellular targeting, and tumor targeting are important for successful therapeutic applications. Another work used gold nanorods and nanoprisms with similar surface plasmon resonances in the near-infrared region. Photothermal therapy was studied in two biological systems, melanoma cells and the small invertebrate Hydra Vulgaris. Results showed a diverse outcome of nanorods and nanoprisms from necrosis to programmed cell death (apoptosis and necroptosis).[14] Similarly, an in depth mechanism of therapy was studied using surface-enhanced Raman spectroscopy (SERS) and mass spectroscopy. Increased levels of phenylalanine, its derivatives, and phenylalanine-containing peptides were observed. It was concluded this resulted due to perturbed phenylalanine metabolism and activation of certain apoptosis pathways.[15]

7.4 FUNCTIONAL GOLD NANOROD-BASED PHOTOTHERMAL THERAPY

Nanomaterials have shown great potential in cancer therapy, but the phenomenon of poor tumor recognition without cellular organelle accumulation usually leads to reduced therapeutic effects and increased side effects. Thus, a variety of functional gold nanorods have been designed for targeting of specific cell/tissue followed by photothermal therapy. For example, silica-coated gold nanorod was prepared with improved biocompatibility. Next, folic acid molecules were conjugated at their surface and used for selective cancer cell targeting, enhanced radiation therapy, and photothermal therapy.[16] These particles also acted as contrast agents for *in vivo* X-ray and computed tomography (CT) imaging. In another work, a sialic acid-imprinted gold nanorod was designed for targeted cancer photothermal therapy.[17] The nanoparticles exhibited good specificity, enabling the killing of cancer cells without damaging healthy cells. In another work, a peptide-terminated gold nanorod was designed for mitochondrial targeting. This nanoplatform promoted effective near-infrared light-triggered subcellular hyperthermia *in vitro* and exhibited excellent tumor ablation ability *in vivo*.[18]

Most cancer-related deaths come from metastasis where a group of cancer cells migrates together via enhanced cytoskeleton filament contraction and coordination with neighboring cells by the cell junction proteins. Although nanoparticles could inhibit individual cancer cell migration, the effect of nanoparticle treatment on collective cell migration has not been explored. One work found that when exposed to light, gold nanorods can inhibit the collective migration of cancer cells by altering the actin filaments and cell junctions via triggered phosphorylation of essential proteins.[19]Another work showed that gold nanorods can open ion channels under light irradiation and regulate apoptotic protein expression. This leads to enhanced cell apoptosis, which might be an effective approach to cancer therapy.[20]

Bacterial biofilms associated with orthopedic implants are notorious cell associations of pathogens that show resistance against antibiotic medications and the host immune response. The presence of hard-to-penetrate extracellular polymeric substances in the biofilm provides a protective shield against the different modes of action of conventional antimicrobial agents. A treatment approach was developed based on a thermoresponsive hydrogel nanocomposite system, containing amino acids and engineered gold nanorods.[21] The composite underwent sol-to-gel transformation at physiological temperatures for site-specific sustained drug release. The released amino acid mixtures facilitate biofilm disruption and gold nanorods offer photothermal treatment that completely eradicates any remaining bacteria cells.

Delivery of chemotherapeutic agents after encapsulation in nanocarriers decreases the dosing to healthy tissues and accumulates in tumors due to the enhanced permeability and retention effect. However, delivering drugs to tumor cells is limited potentially by the rate of release from the carriers. Gold nanorod-based thermosensitive liposomal nanocarriers were designed that maintain the encapsulation of the doxorubicin payload and the option of remote triggered release of doxorubicin under near-infrared illumination.[22] Mouse tumor model results show a significant increase in efficacy when compared to control liposomes.

In another work, polydopamine-coated gold nanorods were incorporated into a thermosensitive injectable hydrogel that was composed of chitosan derivative and dopamine-modified alginate.[23] The sol–gel transition temperature of the composite hydrogel was adjusted to the body temperature (at around 37°C). The hydrogel was quickly and locally heated to over 50°C with NIR irradiation at 808 nm of wavelength. The *in vivo* antitumor test demonstrated the suppression to tumor growth using the hydrogel under multiple photothermal therapy (Figure 7.1).

Despite the advancements in photodynamic therapy and photothermal therapy, the use of a combined photodynamic and photothermal cancer therapy under a single near-infrared irradiation remains limited. A gold nanorod-based sub-100 nm nanosystem was designed for simultaneous near-infrared photodynamic and photothermal cancer therapy.[24] The gold nanorod was terminated with naphthalocyanine and stabilized with alkylthiol-linked polyethylene glycol. The loading of the naphthalocyanine-derivative offered

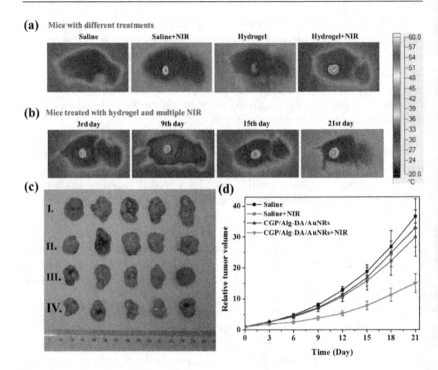

FIGURE 7.1 *In vivo* antitumor activity of composite hydrogel in HepG2-bearing Balb/c mice. Infrared thermal images of (a) mice with various treatments and (b) the mice injected with chitosan-alginate-dopamine-nanorod (CGP/Alg-DA/AuNR) hydrogel under multiple times of NIR irradiation. (c) Photograph of the excised tumors after 21 days ((I) saline, (II) saline with NIR, (III) CGP/Alg-DA/AuNRs, and (IV) CGP/Alg-DA/AuNRs with NIR) and (d) changes in relative tumor volume of HepG2-bearing Balb/c mice upon various treatments, calculated from panel (c). Reprinted with permission from Zeng, J. et al. 2019. Injectable and near-infrared-responsive hydrogels encapsulating dopamine-stabilized gold nanorods with long photothermal activity controlled for tumor therapy. *Biomacromolecules*, 20, 3375–3384. Copyright 2019 American Chemical Society.

a near-infrared photodynamic therapy and gold nanorod offers photothermal therapy. The resulting composite exhibits superior efficacy in cancer cell destruction as compared to photodynamic therapy and photothermal therapy alone. In another work, gold nanorod is conjugated with rose bengal where rose bengal is used to generate singlet oxygen under green light illumination and gold nanorod is used for photothermal effect under near-infrared irradiation. *In vitro* experiments show that reactive oxygen species generated by green light and hyperthermia produced by near-infrared light constitute two different mechanisms for cancer cell death.[25] Compared to single photodynamic therapy or photothermal therapy, the combined effect provided better thera-peutic effects against oral cancer.

The combination of photothermal therapy and chemotherapy as a promis-ing strategy has drawn extensive attention due to overcoming the limitations of conventional treatments in tumor therapy. Gold nanorod-based nanoplatforms are designed via doxorubicin loading to polydopamine-coated gold nanorods and used for tumor metastasis inhibition and multifunctional drug delivery.[26] Gold nanorod offers photothermal therapy as well as the light-induced release of doxorubicin to the tumor microenvironment. Results showed enhanced therapeutic performance of tumors (Figure 7.2).

A combination of photothermal therapy and chemotherapy has been demonstrated as a more effective treatment for complex diseases.[27] A drug delivery system was designed that is composed of mesoporous silica-coated gold nanorod; the mesoporous silica shell is loaded with methotrexate. Folate modification on the surface offers specific targeting to activated macrophages in rheumatoid arthritis. Under 808 nm laser irradiation, this particle kills mac-rophages via local hyperthermia. In addition, internal heating caused the rapid release of methotrexate for localized synergistic photothermal and chemother-apy. In rats with adjuvant-induced arthritis, synergistic treatment inhibits the progression of rheumatoid arthritis.

A near-infrared femtosecond laser is applied for the destruction of pre-formed amyloid-beta (Aβ) fibrils that are adsorbed at the surface of gold nanorods.[28] Alzheimer's disease is associated with the aggregation of the Aβ peptides into toxic aggregates. Although inhibition of Aβ fibrillation has been extensively studied, disintegration/elimination of fibrils has rarely been reported. Results have demonstrated that 800 nm laser irradiation can locally trigger the explosion of nanorods due to the strong localized surface plasmon resonance effect. As a result, the majority of Aβ fibrils are destroyed into small fragments. Furthermore, the laser-induced destruction of fibrils by Au nano-spheres is also investigated, which reveals that most of the Aβ fibrils remain stable under the surface explosion of spherical nanoparticles.

7.5 CONCLUSION

In summary, gold nanorod is shown to have high potential in photothermal therapy. The tunable plasmon band from a visible to near-infrared region, col-loidal nature, and small size of gold nanorods offers unique advantages. In particular, appropriate size and functionalization allow nanorods to be used for cell/tumor targeting followed by photothermal therapy. Ongoing research

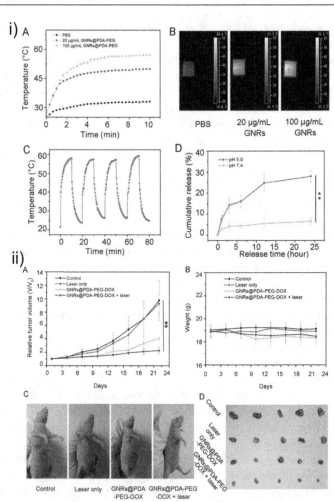

FIGURE 7.2 i) Temperature-rising assessment and drug release from gold nanorod-polydopamine (GNRs@PDA)-based nanocomposites. (A) Temperature-rising profiles of PEGylated GNRs@PDA–PEG under NIR laser irradiation; (B) pictures of control and GNRs@PDA–PEG (20 and 100 µg/mL) obtained from an IR camera; (C) temperature-rising assessment of doxorubicin loaded nanocomposite under NIR laser irradiation over four lasers on/off cycles; and (D) release profiles of DOX under pH 7.4 and pH 5.0. ($n = 3$, **$p < 0.01$). ii) Photothermal/chemo combinatorial tumor therapy *in vivo* for HeLa xenograft tumor mice. (A) Tumor growth curves; (B) average weights of tumor-bearing mice; (C) representative photographs of xenograft-bearing mice from different treatment groups; (D) photographs of tumors in different treatment groups. Reprinted with permission from Li, B. et al. 2019. Gold nanorods-based smart nanoplatforms for synergic thermotherapy and chemotherapy of tumor metastasis. *ACS Applied Materials and Interfaces*, 11, 7800–7811. Copyright 2019 American Chemical Society.

in this area includes subcellular targeting, enhanced *in vivo* targeting, and enhanced light-to-heat conversion efficiency that can further improve photothermal performance. In addition, further understanding of the mechanism of the photothermal effect on living cells can enhance the application performance.

REFERENCES

1. Field, S. B. and Bleehen, N. M. 1979. Hyperthermia in the treatment of cancer. *Cancer Treatment Reviews*, 6, 63–94.
2. Abadeer, N. S. and Murphy, C. J. 2016. Recent progress in cancer thermal therapy using gold nanoparticles. *The Journal of Physical Chemistry C*, 120, 4691–4716.
3. Hirsch, L. R., Stafford, R. J., Bankson, J. A., Sershen, S. R., Rivera, B., Price, R. E., Hazle, J. D., Halas, N. J. and West, J. L. 2003. Nanoshell mediated near-infrared thermal therapy of tumors under magnetic resonance guidance. *The Proceedings of the National Academy of Sciences U. S. A.*, 100, 13549–13554.
4. El-Sayed, I. H., Huang, X. and El-Sayed, M. A. 2006. Selective laser photothermal therapy of epithelial carcinoma using anti-EGFR antibody conjugated gold nanoparticles. *Cancer Letters*, 239, 129–135.
5. Link, S. and El-Sayed, M. A. 2000. Shape and size dependence of radiative, non-radiative and photothermal properties of gold nanocrystals. *International Reviews in Physical Chemistry*, 19, 409–453.
6. Huang, H.-C., Rege, K. and Heys, J. J. 2010. Spatiotemporal temperature distribution and cancer cell death in response to extracellular hyperthermia induced by gold nanorods. *ACS Nano*, 4, 2892–2900.
7. Jang, B. Kim, Y. S. and Choi, Y. 2011. Effect of gold nanorod concentration on the depth-related temperature increase during hyperthermic ablation. *Small*, 7, 265–270.
8. Chen, H., Shao, L., Ming, T., Sun, Z., Zhao, C., Yang, B. and Wang, J. 2010. Understanding the photothermal conversion efficiency of gold nanocrystals. *Small*, 6, 2272–2280.
9. Richardson, H. H. Carlson, M. T. Tandler, P. J. Hernandez, P. and Govorov, A. O. 2009. Experimental and theoretical studies of light-to heat conversion and collective heating effects in metal nanoparticle solutions. *Nano Letters*, 9, 1139–1146.
10. Meyer, S. M., Pettine, J., Nesbitt, D. J. and Murphy, C. J. 2021. Size effects in gold nanorod light-to-heat conversion under femtosecond illumination. *The Journal of Physical Chemistry C*, 125, 16268–16278.
11. Melamed, J. R., Edelstein, R. S. and Day, E. S. 2015. Elucidating the fundamental mechanisms of cell death triggered by photothermal therapy. *ACS Nano*, 9, 6–11.
12. Tong, L., Zhao, Y., Huff, T. B., Hansen, M. N., Wei, A. and Cheng, J.-X. 2007. Gold nanorods mediate tumor cell death by compromising membrane integrity. *Advanced Materials*, 19, 3136–3141.

13. Chen, C.-L., Kuo, L.-R., Chang, C.-L., Hwu, Y.-K., Huang, C.- K., Lee, S.-Y., Chen, K., Lin, S.-J., Huang, J.-D. and Chen, Y.-Y. 2010. In situ real-time investigation of cancer cell photothermolysis mediated by excited gold nanorod surface plasmons. *Biomaterials*, 31, 4104–4112.

14. Moros, M., Lewinska, A., Merola, F., Ferraro, P., Wnuk, M., Tino, A. and Tortiglione, C. 2020. Gold nanorods and nanoprisms mediate different photothermal cell death mechanisms in vitro and in vivo. *ACS Applied Materials and Interfaces*, 12, 13718–13730.

15. Ali, M. R. K., Wu, Y., Han, T., Zang, X., Xiao, H., Tang, Y., Wu, R., Fernandez, F. M. and El-Sayed, M. A. 2016. Simultaneous time-dependent surface-enhanced Raman spectroscopy, metabolomics, and proteomics reveal cancer cell death mechanisms associated with gold nanorod photothermal therapy. *Journal of the American Chemical Society*, 138, 15434–15442.

16. Huanga, P., Baob, L., Zhanga, C., Lina, J., Luoa, T., Yanga, D., Hea, M., Lia, Z., Gaoa, G., Gaoa, B., Fuc, S. and Cui, D. 2011. Folic acid-conjugated silica-modified gold nanorods for X-ray/CT imaging-guided dual-mode radiation and photo-thermal therapy. *Biomaterials*, 32, 9796e9809.

17. Yin, D., Li, X., Ma, Y. and Liu, Z. 2017. Targeted cancer imaging and photothermal therapy via monosaccharide-imprinted gold nanorods. *Chemical Communications*, 53, 6716.

18. Jin, X., Yang, H., Mao, Z. and Wang, B. 2021. Cathepsin B-responsive multifunctional peptide conjugated gold nanorods for mitochondrial targeting and precise photothermal cancer therapy. *Journal of Colloid and Interface Science*, 601, 714–726.

19. Wu, Y., Ali, M. R. K., Dong, B., Han, T., Chen, K., Chen, J., Tang, Y., Fang, N., Wang, F. and El-Sayed, M. A. 2018. Gold nanorod photothermal therapy alters cell junctions and actin network in inhibiting cancer cell collective migration. *ACS Nano*, 12, 9279–9290.

20. Song, J., Pan, J.-B., Zhao, W., Chen, H.-Y. and Xu, J.-J. 2020. Gold nanorod-assisted near-infrared light-mediated regulation of membrane ion channels activates apoptotic pathways. *Chemical Communications*, 56, 6118–6121.

21. Wickramasinghe, S., Ju, M., Milbrandt, N. B., Tsai, Y. H., Navarreto-Lugo, M., Visperas, A., Klika, A., Barsoum, W., Higuera-Rueda, C. A. and Samia, A. C. S. 2020. Photoactivated gold nanorod hydrogel composite containing Damino acids for the complete eradication of bacterial biofilms on metal alloy implant materials. *ACS Applied Nano Materials*, 3, 5862–5873.

22. Agarwal, A., Mackey, M. A., El-Sayed, M. A. and Bellamkonda, R. V. 2011. Remote triggered release of doxorubicin in tumors by synergistic application of thermosensitive liposomes and gold nanorods. *ACS Nano*, 5, 4919–4926.

23. Zeng, J., Shi, D., Gu, Y., Kaneko, T., Zhang, L., Zhang, H., Kaneko, D. and Chen, M. 2019. Injectable and near-infrared-responsive hydrogels encapsulating dopamine-stabilized gold nanorods with long photothermal activity controlled for tumor therapy. *Biomacromolecules*, 20, 3375–3384.

24. Bhana, S., O'Connor, R., Johnson, J., Ziebarth, J. D., Henderson, L. and Huang, X. 2016. Photosensitizer-loaded gold nanorods for near infrared photodynamic and photothermal cancer therapy. *Journal of Colloid and Interface Science*, 469, 8–16.

25. Wang, B., Wang, J.-H., Liu, Q., Huang, H., Chen, M., Li, K., Li, C., Yu, X.-F. and Chu, P. K. 2014. Rose-bengal-conjugated gold nanorods for in vivo photodynamic and photothermal oral cancer therapies. *Biomaterials*, 35, 1954–1966.
26. Li, B., Wang, Y. and He, J. 2019. Gold nanorods-based smart nanoplatforms for synergic thermotherapy and chemotherapy of tumor metastasis. *ACS Applied Materials and Interfaces*, 11, 7800–7811.
27. Li, X., Hou, Y., Meng, X., Li, G., Xu, F., Teng, L., Sun, F. and Li, Y. 2021. Folate receptor-targeting mesoporous silicacoated gold nanorod nanoparticles for the synergistic photothermal therapy and chemotherapy of rheumatoid arthritis. *RSC Advances*, 11, 3567–3574.
28. Lin, D., He, R., Li, S., Xu, Y., Wang, J., Wei, G., Ji, M. and Yang, X. 2016. Highly efficient destruction of amyloidβ fibrils by femtosecond laser-induced nanoexplosion of gold nanorods. *ACS Chemical Neuroscience*, 7, 1728–1736.

Gold Nanorod as Plasmonic Photocatalyst

8

8.1 INTRODUCTION

The plasmonic excitation of gold nanorods gives rise to a collection of optical and electronic effects and induces intense local electric field enhancement near the nanorod surface. The surface plasmons in an excited state can decay non-radiatively by transferring the energy to the charge carriers such as hot electrons and holes.[1] The energetic charge carriers are transferred to the surroundings or relaxed by local heating. On the other hand, the intense local electric field near the nanorod surface can interact with nearby semiconductors or molecules.

Hot electron generation and transfer is the most studied effect involved in gold nanorod-mediated photocatalysis.[2] With the absorption of a photon, free electrons of nanorod can be elevated from the Fermi level to a higher energy level. This high-energy electron (i.e., hot electron) will lose energy through electron–electron scattering in about 10–100 femtoseconds. During this excitation and energy redistribution process, a portion of hot electrons may be transferred to a nearby acceptor, which can be a metal, semiconductor, or reactant molecule. When the acceptor is a semiconductor, the excited hot electrons are injected into the conduction band, and thus the nanorod acts as a photosensitizer.

In order to have more effective plasmonic photocatalysis, various nanocomposites and types of nanostructures have been designed.[2] Other components of designed nanocomposites include metal nanoparticles, semiconductor nanoparticles, and small molecules. The designed nanostructures include core-shell structures, Janus structures, and supported structures. They are applied to various photocatalytic reactions, water splitting and hydrogen evolution, toxic organic degradation, and reactive oxygen species generation-based therapy.

DOI: 10.1201/9781003245339-8

8.2 GOLD NANOROD-BASED PHOTOCATALYSIS MECHANISM

Plasmonic photocatalysis is accompanied by three different catalytic processes: charge transfer-based photoredox catalysis, photo-electrochemical catalysis, and photo-thermal catalysis.[2] In particular, charge transfer-based photoredox catalysis is probably the most common one. This situation occurs for semiconductor-based composite with gold nanorods. When gold nanorods are coupled with semiconductors, two possible phenomena may occur at the hetero-interface under light irradiation: surface plasmon excitation and Schottky barrier formation. The surface plasmon excitation promotes redox catalysis in three ways: hot electron generation and transfer, photothermal conversion, and local electric field enhancement. The Schottky barrier can improve the catalytic activity by modifying the lifetime of electrons.[3] The direction of charge flow would depend on the nature of the composite and modes of photoexcitation. For example, exposing TiO_2–gold nanorod composite to visible light excites the nanorod plasmon, injects hot electrons to TiO_2, and initiates the catalytic reactions.[4-6] Alternatively, if light is used to excite the semiconductor, the excited electrons flow from the conduction band of the semiconductor to the gold nanorod where the nanorod acts as an electron sink which results in redox reactions.[3] Another possibility is that both nanorods and semiconductors are photoexcited. The electron transfer can occur in both ways or becomes more complex.[7] In conclusion, gold nanorod serves as an optical antenna and provides energetic hot electrons. This enhances the local electric field which can assist in generating and separating charge carriers.

Photo-electrochemical catalysis utilizes light energy for electrochemical catalysis.[8, 9] This approach has promise for the conversion of solar energy to chemical energy.[10] A typical design is composed of one photoactive material or photoelectrode and a metal counter electrode immersed in an electrolyte and connected by an external electric wire. The early example of a gold nanorod-based photoelectrode is composed of iron oxide nanorod arrays on a gold nanorod substrate.[11] The composite electrode shows high photocurrent density ($8\ mA\ cm^{-2}$) under light irradiation. Gold nanorod-based other photoelectrodes utilize water oxidation using NIR light-responsive gold nanorod.[12] Under visible light irradiation, O_2 is produced at the anode as a primary product, while H_2O_2 is produced under NIR light. Gold nanorod-based TiO_2–Pt Janus particles are designed for photocatalytic H_2 evolution.[13] With methanol as the sacrificial electron donor, the device can operate for more than 200h with a H_2 production rate of 2.3 mol/h. In general, the photo-electrochemical configuration provides greater flexibility in designing and selecting photoactive materials. However, this approach needs to be extended to different systems such as applications in N_2 and CO_2 reduction, organic synthesis, and biomass oxidation.

Photothermal catalysis is similar to conventional thermal catalytic reactions, except that heating is induced by plasmonic nanoparticles.[1, 2] Plasmonic gold nanorod-based composites can effectively channel the flux of photons into heat to drive the catalytic reactions. In the non-irradiative surface plasmon decay, electrons in gold nanorod with a non-thermal electron distribution will undergo thermalization. Then, cooling of the hot electrons via phonon modes increases the temperature of the nanorod and local environment. This enhances the reaction speed. For example, Pd nanoparticle deposited gold nanorod has been designed for efficient Suzuki coupling reaction under light irradiation.[14] The integration of plasmonic nanorod with catalytic Pd enables efficient Vis–NIR light harvesting and the yield of the Suzuki coupling reaction by Pd is increased two times via local heating. In another work, Pd-tipped gold nanorod is designed for efficient plasmon-enhanced catalytic formic acid dehydrogenation under low temperatures (5°C)[15] (Figure 8.1). It has been shown that with light irradiation, the H_2 evolution rate increases nearly 15 times compared with that under the dark at 40°C. It is important to mention that photothermal catalysis is often accompanied by photo-electrochemical and photoredox catalysis and complicates the reaction mechanism.

8.3 PLASMONIC PHOTOCATALYTIC APPLICATION OF GOLD NANOROD-BASED DIFFERENT NANOCOMPOSITES

The Plasmon-induced photocatalytic property of gold nanorods has been used for different applications. They include photocatalytic water splitting and hydrogen evolution, toxic organic degradation, and reactive oxygen species generation-based therapeutic applications. Here we will briefly cover these applications with some representative examples.

Gold nanorods are known to exhibit strong longitudinal surface plasmon bands, leading to very efficient energy conversion under near-IR excitation. Near-IR ultrashort-pulsed irradiation on gold nanorods is shown to induce the production of reactive oxygen species (ROS) over a few micrometers in their aqueous environment, particularly hydroxyl radicals.[16] The variation of the excitation wave polarization angle relative to the nanorod orientation reveals a nonlinear polarization-controlled action combined with a three-photon

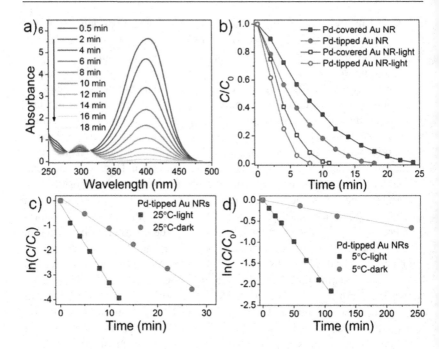

FIGURE 8.1   (a) Time-dependent UV-visible absorption spectra of the reduction of 4-nitrophenol (4-NP) by NaBH$_4$ in the presence of Pd-tipped gold nanorods (Au NRs) (0.008 mg) in the dark at 25°C. (b) Time-dependent profiles for this catalytic reaction with or without light irradiation (λ > 520 nm) at 25°C. (c and d) Kinetic linear fitting curves for the reduction of 4-NP over Pd-tipped Au NRs at 25 and 5°C. Reprinted with permission from Zheng, Z. et al. 2015. Plasmon-enhanced formic acid dehydrogenation using anisotropic Pd–Au nanorods studied at the single-particle level. *Journal of the American Chemical Society*, 137, 948–957. Copyright 2015 American Chemical Society.

luminescence of the nanorod. The longitudinal plasmon enhances the local electromagnetic field and therefore controls both luminescence and ROS production through multiphoton excitation of electrons in gold and water followed by hot plasma formation in water. These results offer new insights into the potential use of nanorods as a replacement for photosensitizers in photodynamic cancer therapy or in combinations of therapeutic and imaging methods.

In another work, gold nanorods and gold nanocluster-based hybrid heterostructures were designed that exhibit both localized surface plasmon resonance properties and peroxidase-like activity.[17] It was found that the catalytic activity of the heterostructure is remarkably enhanced by visible to near-infrared (NIR) light excitation. The hybrid structure is functionalized with antibodies for colorimetric assay of prostate-specific antigens. Here the peroxidase-like activity of heterostructure is sensitive to target molecule binding that controls the photon–plasmon coupling. This sensor is practically applied to detect prostate-specific antigen levels in prostate cancer serum samples.

Palladium nanoparticles have been reported to have various nanozyme activities and exhibit promising potential for biomedical applications. However, as Pd is a poor plasmonic metal, little attention has been paid to its plasmon-regulated nanozyme activity. Pd-coated Au nanorods were designed where the nanorod acts as a strong plasmonic core, and the Pd shell acts as a nanozyme.[18] The obtained hybrid nanoparticle showed tunable plasmon bands in the near-infrared (NIR) light irradiation-based catalysis by Pd. Figure 8.2 shows the

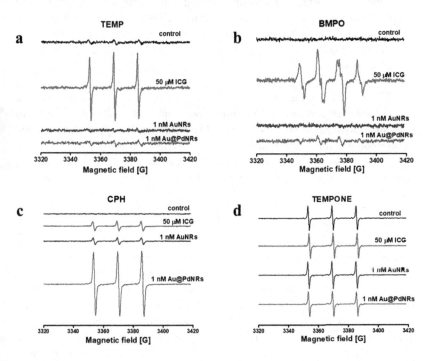

FIGURE 8.2 (a–d) Characterizing intermediates of O_2 activation on gold nanorod surface using ESR spectral method. ESR spectra of (a) 2,2,6,6-tetramethyl-4-piperidine (TEMP), (b) 5-tert-Butoxycarbonyl-5-methyl-1-pyrroline N-oxide (BMPO), (c) 1-hydroxy-3-carboxy2,2,5,5-tetramethylpyrrolidine hydrochloride (CPH), and (d) TEMPONE in the presence of Pd coated gold nanorod (Au@PdNRs), gold nanorod (AuNR), or indocyanine green (ICG). All samples were placed in a 96-well plate and irradiated with 1 W/cm² 808 nm laser for 5 min before ESR measurements. Reprinted with permission from Fan, H. et al. 2019. Plasmon-enhanced oxidase-like activity and cellular effect of Pd coated gold nanorods. *ACS Applied Materials and Interfaces*, 11, 45416–45426. Copyright 2019 American Chemical Society.

evidence of reactive oxygen species generation by gold nanorod-based photocatalysis using Pd-coated Au nanorod. The plasmon-enhanced oxidase-like activity has been observed under NIR light irradiation and is mainly ascribed to the local photothermal effect. It is used to induce cytotoxicity and depolarize mitochondrial membrane potential.

Plasmonic metal/semiconductor heterostructures show promise for visible-light-driven photocatalytic water splitting and hydrogen evolution. Gold nanorods semi-coated with TiO_2 are expected to be ideally structured systems for hydrogen evolution.[19] Gold nanorod/TiO_2 nanodumbbells with spatially separated Au/TiO_2 regions are designed and used for plasmon-enhanced H_2 evolution under visible and near-infrared light. Similarly, gold nanorod-based other nanocomposites are designed for solar energy conversion and sunlight-based water splitting. They include iron oxide nanorod arrays on gold nanorod substrate for photocurrent generation,[11] gold nanorod-based water oxidation, and O_2 production at anode using NIR light[12] and photocatalytic H_2 evolution from gold nanorod-based TiO_2-Pt Janus particles.[13]

In another work, the end-to-end assembly of gold nanorods is designed via the H-bonded interaction of para-aminothiophenol that is selectively bound to the longitudinal ends of the nanorods.[20] The plasmonic hot spots generated due to nanorod assembly are used as a catalytic center for a photochemical coupling reaction of aminothiophenol to di-mercaptoazobenzene. The coupling reaction was carried out at the hot spot of the assembled nanorods upon surface plasmon excitation with a 1.6-fold faster reaction rate.

The development of sustainable methods for the degradation of pollutants in water is an ongoing critical challenge. Anthropogenic organic micropollutants, such as pharmaceuticals present in our water supplies in trace quantities, are currently not remediated by conventional treatment processes. In this regard a method was created for the oxidative degradation of organic micropollutants using specially designed nanoparticles and a visible wavelength of sunlight.[21] Gold "Janus" nanorods, partially coated with silica, were synthesized where silica was used to enhance their colloidal stability in aqueous solutions while also maintaining a partially uncoated Au surface to facilitate photocatalysis. Au structures are dispersed in an aqueous solution containing peroxydisulfate where oxidative degradation of both simulant and actual organic micropollutants was observed. Photothermal heating, light-induced hot electron-driven charge transfer, and direct electron shuttling under dark conditions all contribute to the observed oxidation chemistry. This work not only provides an ideal platform for studying plasmonic photochemistry in aqueous medium but also opens the door for nanoengineered, solar-based methods to remediate recalcitrant micropollutants in water supplies. Other examples of enhanced chemical reactions include Pd nanoparticle deposited gold nanorods for 2-times enhanced Suzuki coupling reactions under Vis-NIR light irradiation,[14] Pd-tipped gold nanorods for plasmon-enhanced catalytic formic acid dehydrogenation, and 15-times enhanced H_2 evolution rate.[15]

8.4 CONCLUSION

In summary, the tunable plasmon bands of gold nanorods from a visible to near-infrared region have been shown to have high potential in plasmon-based photochemical reactions. In particular, gold nanorod-based various nanocomposites are designed for light-enhanced chemical reaction, water splitting, hydrogen evolution, toxic organic degradation, and reactive oxygen species generation. Ongoing research in this area includes improved nanocomposite design for enhanced plasmonic photocatalysis, understanding the mechanism of plasmonic photocatalysis, and more effective utilization in chemical and biomedical science.

REFERENCES

1. Hou, W. And Cronin, S. B. 2013. A review of surface plasmon resonance-enhanced photocatalysis. *Advanced Functional Materials*, 23, 1612–1619.
2. Hana, C., Qi, M.-Y., Tanga, Z.-R., Gong, J. And Xu, Y.-J. 2019. Gold nanorods-based hybrids with tailored structures for photoredox catalysis: Fundamental science, materials design and applications. *Nano Today*, 27, 48–72.
3. Zhang, P., Wang, T. And Gong, J. 2015. Mechanistic understanding of the plasmonic enhancement for solar water splitting. *Advanced Materials*, 27, 5328–5342.
4. Erwin, W. R., Coppola, A., Zarick, H. F., Arora, P., Miller, K. J. And Bardhan, R. 2014. Plasmon enhanced water splitting mediated by hybrid bimetallic Au–Ag core–shell nanostructures. *Nanoscale*, 6, 12626–12634.
5. Li, B., Gu, T., Ming, T., Wang, J., Wang, P., Wang, J. And Yu, J. C. 2014. Gold core@ceria shell nanostructures for plasmon-enhanced catalytic reactions under visible light. *ACS Nano*, 8, 8152–8162.
6. Zheng, Z., Tachikawa, T. And Majima, T. 2015. Plasmon-induced spatial electron transfer between single Au nanorods and ALD-coated TiO_2: dependence on TiO_2 thickness. *Chemical Communications*, 51, 14373–14376.
7. Ha, J. W., Ruberu, T. P. A., Han, R.., Dong, B., Vela, J. And Fang, N. 2014. Super-resolution mapping of photogenerated electron and hole separation in single metal–semiconductor nanocatalysts. *Journal of the American Chemical Society*, 136, 1398–1408.

8. Lee, J. B., Choi, S., Kim, J. And Nam, Y. S. 2017. Plasmonically-assisted nanoarchitectures for solar water splitting: Obstacles and breakthroughs. *Nano Today*, 16, 61–81.

9. Li, C., Wang, T., Zhao, Z.-J., Yang, W., Li, J.-F., Li, A., Yang, Z., Ozin, G. A. And Gong, J. 2018. Promoted fixation of molecular nitrogen with surface oxygen vacancies on plasmon-enhanced TiO_2 photoelectrodes. *Angewandte Chemie International Edition*, 57, 5278–5282.

10. Ghobadi, T. G. U., Ghobadi, A., Ozbay, E. And Karadas, F. 2018. Strategies for plasmonic hot-electron-driven photoelectrochemical water splitting. *ChemPhotoChem*, 2, 161–182.

11. Mao, A., Han, G. Y. And Park, J. H. 2010. Synthesis and photoelectrochemical cell properties of vertically grown α-Fe_2O_3 nanorod arrays on a gold nanorod substrate. *Journal of Materials Chemistry*, 20, 2247–2250.

12. Nishijima, Y., Ueno, K., Kotake, Y., Murakoshi, K., Inoue, H. And Misawa, H. 2012. Near-infrared plasmon-assisted water oxidation. *The Journal of Physical Chemistry Letters*, 3, 1248–1252.

13. You, B., Liu, X., Jiang, N. And Sun, Y. 2016. A general strategy for decoupled hydrogen production from water splitting by integrating oxidative biomass valorization. *Journal of the American Chemical Society*, 138, 13639–13646.

14. Wang, F., Li, C., Chen, H., Jiang, R., Sun, L.-D., Li, Q., Wang, J., Yu, J. C. And Yan, C.-H. 2013. Plasmonic harvesting of light energy for Suzuki coupling reactions. *Journal of the American Chemical Society*, 135, 5588–5601.

15. Zheng, Z., Tachikawa, T. And Majima, T. 2014. Plasmon-enhanced formic acid dehydrogenation using anisotropic Pd–Au nanorods studied at the single-particle level. *Journal of the American Chemical Society*, 137, 948–957.

16. Labouret, T., Audibert, J.-F., Pansu, R. B. And Palpant, B. 2015. Plasmon-assisted production of reactive oxygen species by single gold nanorods. *Small*, 11, 4475–4479.

17. Tan, F., Yang, Y., Xie, X., Wang, L., Deng, K., Xia, X., Yanga X. And Huang, H. 2018. Prompting peroxidase-like activity of gold nanorod composites by localized surface plasmon resonance for fast colorimetric detection of prostate specific antigen. *Analyst*, 143, 5038–5045.

18. Fan, H., Li, Y., Liu, J., Cai, R., Gao, X., Zhang, H., Ji, Y., Nie, G. And Wu, X. 2019. Plasmon-enhanced oxidase-like activity and cellular effect of Pd coated gold nanorods. *ACS Applied Materials and Interfaces*, 11, 45416–45426.

19. Wu, B., Liu, D., Mubeen, S., Chuong, T. T., Moskovits, M. And Stucky, G. D. 2016. Anisotropic growth of TiO_2 onto gold nanorods for plasmon enhanced hydrogen production from water reduction. *Journal of the American Chemical Society*, 138, 1114–1117.

20. Pal, S., Dutta, A., Paul, M. And Chattopadhyay, A. 2020. Plasmon-enhanced chemical reaction at the hot spots of end-to end assembled gold nanorods. *The Journal of Physical Chemistry C*, 124, 3204–3210.

21. Weia, H., Loeba, S. K., Halas, N. J. And Kim, J.-H. 2020. Plasmon-enabled degradation of organic micropollutants in water by visible-light illumination of Janus gold nanorods. *The Proceedings of the National Academy of Sciences U. S. A.*, 117, 15473–15481.

Gold Nanorod in Electrochemical Applications

9

9.1 INTRODUCTION

The electrochemical application of gold nanorods is relatively less popular as compared to their other applications.[1,2] This is particularly because the smaller size and spherical gold nanoparticle can perform similar electrochemical performance. However, there are some advantages of gold nanorods that have been utilized in practical applications. In the electrochemical application, gold nanorods are primarily used for the detection of electrochemicals, and as photo-electrochemical material in solar energy conversion and electrode material in fuel cells. There are three specific advantages of using gold nanorods in these applications.[2] First, an anisotropic shape offers a relatively higher surface with different atomic planes that induce unique electrochemical reactions at the electrode surface. In particular, nanorod offers the generation of enhanced reactive oxygen species, the formation of light-induced electroactive species, and the generation of photocurrents. Second, intense and tunable plasmon bands can be used as photo-electrochemical material. Third, a relatively larger surface area can offer larger space for conjugating affinity biomolecules or other nanoparticles for enhanced electrochemical performance. The electrochemical application of gold nanorods can be divided into two types: the application of gold nanorods in electrochemical detection and the application of gold nanorods in the generation of fuel cells and photocurrents. Here we will summarize these applications along with the design of gold nanorod-based nanocomposites.

DOI: 10.1201/9781003245339-9

9.2 APPLICATION OF GOLD NANOROD IN ELECTROCHEMICAL DETECTION

The electrochemical detection application of gold nanorods is primarily based on their conducting property and capability in producing various electroactive species at the nanorod surface. Thus gold nanorod-based different nanocomposites are designed and used to modify working electrodes. Next, electroactive species are determined via an electrochemical reaction that in turn is linked to the detection of some specific chemicals/biochemicals. For example, horseradish peroxidase functionalized gold nanorods are designed for sensitive electrochemical detection of alpha-fetoprotein antigens.[3] First, horseradish peroxidase and secondary antibody functionalized gold nanorods are designed to label the signal antibodies for sensitive electrochemical measurement of alpha-fetoprotein. The immunosensor was fabricated by assembling carbon nanotubes and gold nanorods and captures antibodies on the glassy carbon electrode. This immunosensor assembly is incubated with alpha-fetoprotein antigen followed by horseradish peroxidase and secondary antibody functionalized gold nanorods. This captures functionalized gold nanorods on the surface of the gold nanorod–carbon nanotube via specific recognition of antigen-antibody. Differential pulse voltammetry has been employed to record the electrochemical signal response of a mixture of hydrogen peroxide (H_2O_2) and 3, 30, 5, 50-tetramethylbenzidine. The signal intensity is linearly related to the concentration of alpha-fetoprotein in the range of 0.1–100 ng/mL. It has been shown that the signal intensity of the immunosensor is increased by 2.6-fold in the presence of nanorod and the reason is that nanorod offers more loading of horseradish peroxidase and results in the increased response signal.

In another example, Ag-coated gold nanorod is incorporated into As/Sb/Bi nanosheets and used as novel electrode materials for a non-enzymatic electrochemical glucose sensor.[4] First, silver-coated gold nanorods are prepared using a seeded-growth method. As/Sb/Bi nanosheets are shear exfoliated separately by using kitchen blenders and used to make nanocomposite. This composite is used as an enzyme-free glucose sensor. The non-enzymatic glucose detection mechanism involves the formation of hydroxide moieties on the surface of the electrode to form an M–OH premonolayer. Adsorption of glucose onto the active sites of the electrodes occurs through a nucleophilic attack by the OH adsorbed on M on the C1 carbon on glucose followed by C1 dehydrogenation. The glucose subsequently electro-oxidizes to produce glucono-d-lactone which then further undergoes oxidation into gluconic acid. A sharp distinguishable electrochemical signal linked to glucose oxidation is produced when an arsenic nanosheet-based nanorod composite is used. This approach has been used for highly sensitive enzyme-less glucose sensors.

9.3 APPLICATION OF GOLD NANOROD IN FUEL CELL AND PHOTOCURRENT GENERATION

One useful electrochemical application of gold nanorods is as fuel cell electrode material. Here the role of gold nanorod is to enhance the electrochemical reaction at the electrode surface. For example, porous Au nanorod has been synthesized via de-alloying of Au–Ag nanorods and then coated with Pt.[5] The catalytic activities of Pt-coated porous Au nanorods with different lengths show that smaller ones are better for use as a cathode material during methanol electro-oxidation. This is principally due to different amounts of carbon monoxide (CO) generated during methanol electro-oxidation and less carbon monoxide is generated for shorter nanorods. In another work, gold nanorod-polyaniline nanocomposites with different levels of gold nanorods were prepared. They were used as anode material in borohydride oxidation reactions and for potential application in direct borohydride-peroxide fuel cells.[6] The nanocomposites showed a power density of 184 Wg-1 in a direct borohydride-peroxide fuel cell.

Another useful electrochemical application of gold nanorods is as electrode material for solar light conversion. Here the role of gold nanorod is to capture light and to enhance the electrochemical reaction at the electrode surface. In particular, gold nanorod has been used in various photo-electrochemical catalysis that utilize light energy for electrochemical catalysis. In a typical design, one photoactive material or photoelectrode and a metal counter electrode are immersed in an electrolyte and connected by an external electric wire. Photocurrents passing through a single photosystem complex located in the vicinity of a single gold nanorod are measured. Simultaneous excitation of the photosystem complex and the gold nanorod longitudinal plasmon mode is shown to enhance the photocurrent by eight times[7] (Figure 9.1). The extent of enhancement depends on the distance between photosystem complexes and gold nanorods and indicates that, apart from the enhancement of absorption, there is an additional enhancement mechanism directly affecting the electron transport process.

In a real example, a gold nanorod-based photoelectrode is composed of iron oxide nanorod arrays on a gold nanorod substrate.[8] The composite

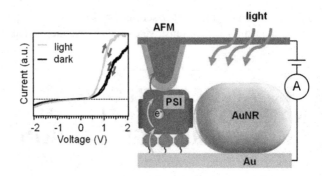

FIGURE 9.1 Right panel: Schematic representation of the sample structure and electrochemical setup showing the location of gold nanorod and photosystem complexes (PSI). Left panel: I–V curves, measured on single PSI complexes located near gold nanorod; light black lines indicate measurements under irradiation with a 635 nm laser, and deep black lines indicate measurements in the dark. Reprinted with permission from Furuya, R., et al. 2021. Enhancement of the photocurrent of a single photosystem I complex by the localized plasmon of a gold nanorod. *Journal of the American Chemical Society*, 143, 13167–13174. Copyright 2021 American Chemical Society.

electrode shows high photocurrent density (8 mA cm^{-2}) under light irradiation. In another example, gold nanorod-based photoelectrode is designed to utilize water oxidation under NIR light[9] where O_2 is produced at the anode as a primary product, while H_2O_2 is produced under NIR light.[9] Similarly, gold nanorod-based TiO_2–Pt Janus particles were designed for photocatalytic H_2 evolution.[10] With methanol as the sacrificial electron donor, the device can operate for more than 200h with a H_2 production rate of 2.3 mol/h. Moreover, gold nanorods semi-coated with TiO_2 are designed for visible-light-driven photocatalytic water splitting and hydrogen evolution.[11]

There is another example where a gold nanorod-based nanocomposite was designed for the electroluminescence sensor of thrombin.[12] First, gold nanorod-based nanocomposite with hemin and graphene is designed via π–π stacking interaction between hemin and graphene and in-situ growth of gold nanorod. Next, two different thrombin aptamers are used to make a sandwich with quantum dots as follows: thrombin aptamer I–thrombin–thrombin aptamer II gold nanorod-graphene-hemin. The interaction between aptamer and thrombin shortens the distance between the quantum dot donor and gold nanorod-graphene-hemin nanocomposite receptor to trigger electrochemiluminescence energy transfer. In addition, hemin is intercalated into an aptamer structure and connected onto a nanorod composite that induces the peroxidase-like catalysis of H_2O_2 to greatly enhance the electroluminescence of luminol. The thrombin can be detected in the range from 100 ng/mL to 0.5 pg/mL.

9.4 CONCLUSION

In summary, gold nanorod has been used as electroactive materials in electrochemical detection, as fuel cell electrode material to enhance the electrochemical reaction at the electrode surface, and as material for solar energy capture and electrochemical conversion at the electrode surface. These electrochemical applications are primarily based on the capability of gold nanorods in producing various electroactive species at the nanorod surface, efficient solar light capturing, and conversion performance. Discussed methods show that gold nanorod has potential in various electrochemical applications.

REFERENCES

1. Lee, J. B., Choi, S., Kim, J. and Nam, Y. S. 2017. Plasmonically-assisted nanoarchitectures for solar water splitting: Obstacles and breakthroughs. *Nano Today*, 16, 61–81.
2. Hana, C., Qi, M.-Y., Tanga, Z.-R., Gong, J. and Xu, Y.-J. 2019. Gold nanorods-based hybrids with tailored structures for photoredox catalysis: Fundamental science, materials design and applications. *Nano Today*, 27, 48–72.
3. Guo, J., Han, X., Wang, J., Zhao, J., Guo, Z. and Zhang, Y. 2015. Horseradish peroxidase functionalized gold nanorods as a label for sensitive electrochemical detection of alpha-fetoprotein antigen. *Analytical Biochemistry*, 491, 58–64.
4. Chia, H. L., Mayorga-Martinez, C. C., Gusma~o, R., Novotny, F., Webster, R. D. and Pumera, M. 2020. A highly sensitive enzyme-less glucose sensor based on pnictogens and silver shell–gold core nanorod composites. *Chemical Communications*, 56, 7909–7912.
5. Yoo, S.-H., Liu, L., Cho, S. H. and Park, S. 2012. Platinum-coated porous gold nanorods in methanol electrooxidation: Dependence of catalytic activity on ligament size. *Chemistry: An Asian Journal*, 7, 2937–2941.
6. Milikic, J., Stamenovic, U., Vodnik, V., Ahrenkiel, S. P. and Sljukic, B. 2019. Gold nanorod-polyaniline composites: Synthesis and evaluation as anode electrocatalysts for direct borohydride fuel cells. *Electrochimica Acta*, 328, 135115.
7. Furuya, R., Omagari, S., Tan, O., Lokstein, H. and Vacha, M. 2021. Enhancement of the photocurrent of a single photosystem I complex by the localized plasmon of a gold nanorod. *Journal of the American Chemical Society*, 143, 13167–13174.
8. Mao, A., Han, G. Y. and Park, J. H. 2010. Synthesis and photoelectrochemical cell properties of vertically grown α-Fe_2O_3 nanorod arrays on a gold nanorod substrate. *Journal of Materials Chemistry*, 20, 2247–2250.

9. Nishijima, Y., Ueno, K., Kotake, Y., Murakoshi, K., Inoue, H. and Misawa, H. 2012. Near-infrared plasmon-assisted water oxidation. *The Journal of Physical Chemistry Letters*, 3, 1248–1252.

10. You, B., Liu, X., Jiang, N. and Sun, Y. 2016. A general strategy for decoupled hydrogen production from water splitting by integrating oxidative biomass valorization. *Journal of the American Chemical Society*, 138, 13639–13646.

11. Wu, B., Liu, D., Mubeen, S., Chuong, T. T., Moskovits, M. and Galen D. Stucky, G. D. 2016. Anisotropic growth of TiO_2 onto gold nanorods for plasmon enhanced hydrogen production from water reduction. *Journal of the American Chemical Society*, 138, 1114–1117.

12. Shao, K., Wang, B., Ye, S., Zuo, Y., Wu, L., Li, Q., Lu, Z., Tan, X. C. and Han, H. 2016. Signal-amplified near-infrared ratiometric electrochemiluminescence aptasensor based on multiple quenching and enhancement effect of graphene/gold nanorods/G-quadruplex. *Analytical Chemistry*, 88, 8179–8187.

Other Applications of Gold Nanorod

10

10.1 INTRODUCTION

In addition to well-known applications of gold nanorods, there are a few other applications that are less explored. However, they are important and deserve separate discussions. These applications include drug delivery carrier, remote delivery of drugs, *in vivo* targeting using the light-responsive property, and as a bioprobe to understand the mechanical property of cellular components.[1–3]

A wide variety of molecular drugs are loaded at the surface of gold nanorods and then used for *in vitro* and *in vivo* delivery. Shape anisotropy of nanorods has three important aspects in those applications. First, it offers different types of interaction with cells and can lead to unique uptake mechanisms.[3] Second, an anisotropic shape with a differential binding surface at the side/tip offers differential loading and functionalization of molecules. Third, nanorod shape has an advantage over similar size spheres in terms of long-term circulation in blood that can offer an advantage for *in vivo* targeting. In addition, the light-responsive property of nanorods can be coupled for additional properties. Here we will summarize those applications along with the respective design and performance.

10.2 GOLD NANOROD AS DELIVERY CARRIER

Nanoparticles have been used as drug carriers to deliver antibiotics directly to the site of infection, reducing any side effects of off-target toxicities.

DOI: 10.1201/9781003245339-10

One of the unique applications of gold nanorods is as a delivery carrier for small molecule drugs. Drugs can be directly adsorbed at the nanorod surface or electrostatically bound to the coating material of the nanorod or conjugated to the coating material. A large surface area can be used to load more drugs and multiple drugs. Moreover, drug release can be controlled using light.

In one example, PEGylated gold nanorods are loaded with Pt(IV) prodrug and used as a delivery system. On entering cells, the Pt(iv) prodrug is reduced by cellular glutathione to the active divalent platinum and released from the nanorod carrier.[4] Results show superior cytotoxicity compared to cisplatin against different types of cancer cells.

In another example, gold nanorod is conjugated to rifampicin and is released after uptake into macrophage cells. Nanorods are actively internalized into macrophages and release drug after uptake, under the safety frame of the host cells (macrophage).[5] Nanorods without rifampicin conjugation do not exhibit antimicrobial activity. Therefore, nanorod could be a promising delivery vehicle for the anti-tuberculosis drug rifampicin for use in tuberculosis therapy.

In another example, protein corona formed around gold nanorod is used for doxorubicin loading.[6] A drug release study conducted over a period of 14 days indicated that both surface functionalization and amount of bound corona had a combinatorial effect on the retention of drug payload. The study showed the application potential of protein-coronated gold nanorods with ideal surface functionalization as a new nanoplatform for drug delivery.

Middle East respiratory syndrome (MERS) coronavirus causes a severe acute respiratory syndrome-like illness with high pathogenicity and mortality due to the lack of effective therapeutics. Currently, only a few antiviral agents are available for the treatment of MERS, but their effects have been greatly impaired by low antiviral activity. A series of peptide inhibitors have been developed to inhibit membrane fusion by coronavirus to the host cells. The inhibitory effect of peptides can be further increased to 10-fold by forming a gold nanorod complex[7] (see Figure 10.1). In addition, gold nanorod enhances the metabolic stability of peptides and, therefore, effectively prevents coronavirus-associated membrane fusion.

Delivery of CRISPR/Cas9 machineries into living cells and tissues is of paramount importance in a wide range of therapeutic applications. Cationic polymer-coated gold nanorods with a high aspect ratio are shown to exhibit excellent capability to mediate internalization, and strong ability to escape endosomes.[8] The intracellular delivery mediated by cationic nanorods enables Cas9-mediated genome editing and dCas9-mediated transcriptional activation, and *in vivo* delivery of CRISPR/Cas9 plasmid-targeting can successfully protect mice from liver fibrosis.

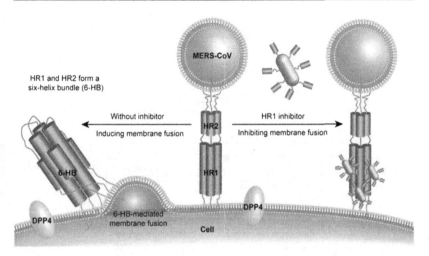

FIGURE 10.1 Schematic diagram of the inhibition mechanism of membrane fusion with peptide (HR1) inhibitors that is functionalized with gold nanorod. HR1 inhibitor can inhibit HR1/HR2 complex (6-HB)-mediated membrane fusion and prevent coronavirus (MERS-CoV) infections. Reprinted with permission from Huang, X., et al. 2019. Novel gold nanorod-based HR1 peptide inhibitor for middle-east respiratory syndrome coronavirus. *ACS Applied Materials and Interfaces*, 11, 19799–19807. Copyright 2019 American Chemical Society.

10.3 GOLD NANOROD-BASED CO-DELIVERY OF MULTIPLE DRUGS

Combination therapy, or the use of multiple drugs, has been proven to be effective for complex diseases. However, differences in chemical properties and pharmacokinetics can be challenging in terms of the loading, delivering, and releasing of multiple drugs. Gold nanorod has been used to load and selectively release different oligonucleotides from their surface.[9] Two different oligonucleotides are loaded on two different nanorods via thiol conjugation. Selective release of oligonucleotide is induced by selective melting of gold nanorods via ultrafast laser irradiation at the nanorods' longitudinal surface plasmon resonance peaks. Excitation at one wavelength could selectively melt one type of gold nanorod and thus selectively release one type of DNA strand. Released oligonucleotides were still functional (see Figure 10.2).

In another work, a multifunctional gold nanorod-based nanocarrier was designed that is capable of co-delivering small interfering RNA and an anticancer

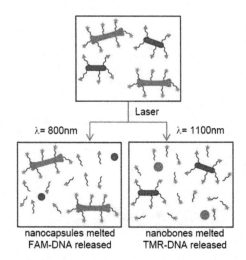

nanocapsules melted
FAM-DNA released

nanobones melted
TMR-DNA released

FIGURE 10.2 Overview of selective release of different DNAs. Laser irradiation of DNA-conjugated nanocapsules (shorter ovals) and nanobones (longer bones) are exposed to λ_{800} irradiation (left), which melts the nanocapsules and selectively releases the conjugated DNA (labeled by FAM (triangles)). Exposure to λ_{1100} irradiation (right) melts the nanobones, selectively releasing the conjugated DNA (labeled by TMR (stars)). Reprinted with permission from Wijaya, A., et al. 2009. Selective release of multiple DNA oligonucleotides from gold nanorods. *ACS Nano*, 3, 80–86. Copyright 2009 American Chemical Society.

drug, specifically to neuroendocrine cancer cells.[10] The nanorod is conjugated with an anticancer drug, psiRNA, via poly-arginine for complexing, and octreotide (a tumor-targeting ligand) for specific targeting of cancer cells with overexpressed somatostatin receptors. Functional nanorods exhibited a much higher performance with significantly higher gene silencing in cancer cells.

10.4 GOLD NANOROD-BASED REMOTE CONTROLLED DRUG DELIVERY

The photodynamic therapy of cancer is associated with the generation of singlet oxygen in the tumor region. However metastatic cancer tumors have hypoxia regions with oxygen deficiency which limits therapeutic effect. Thus generation or delivery of singlet oxygen at remote areas of tumor is challenging. A gold nanorod-based carrier was designed that can deliver singlet oxygen to the cancer hypoxia region.[11] An endoperoxide-conjugated gold nanorod was designed that generated singlet oxygen under 808 nm light irradiation via thermal cycloreversion of endoperoxides. It is shown that the amount of generated singlet oxygen is sufficient for triggering cellular apoptosis.

10.5 LIGHT RESPONSIVE TARGETING BY GOLD NANOROD

Gold nanorods have strong absorbance in the near-infrared region, and the absorbed light energy can be converted to heat. This property has been used for *in vivo* targeting applications. Gold nanorod is coated with thermo-responsive polymer with a phase transition temperature of 34°C. This polymer-coated gold nanorod decreases in size under near-infrared laser irradiation due to a phase transition of the polymer layers.[12] The polymer-coated gold nanorods circulate in blood flow without a phase transition after intravenous injection. However, the irradiation of near-infrared light at a tumor resulted in the gold specifically accumulating in the tumor. This novel accumulation technique and the photothermal effect of the gold nanorod offer a powerful tool for targeted delivery in response to light irradiation.

Transdermal delivery is a useful and attractive method for drug delivery because it is comfortable for patients and has potential benefits. However, transdermal penetration of hydrophilic macromolecular drugs/proteins is challenging. An interesting way to improve the efficiency of the transdermal delivery of hydrophilic macromolecules is to use a solid in-oil dispersion Gold nanorod-based solid-in-oil dispersion was designed which is effective in breaking the skin barrier using the photohermal effect of nanorods.[13] A high-molecular-weight model protein, ovalbumin, is delivered through the skin and induces an immune response in mice by a solid-in-oil dispersion of gold nanorods after irradiation with near-infrared light.

10.6 OTHER APPLICATIONS OF GOLD NANOROD

There are a few other applications of gold nanorods that deserve mention. They use the photothermal property and anisotropic property of nanorods. In one such work, gold nanorod orientation and rotational dynamics are used to understand cellular mechanics.[14] The in-plane rotation and out-of-plane tilting, as well as other more complex rotational patterns, are imaged at a resolution of 5 ms for a single gold nanorod in a live cell. The unique capabilities of visualizing rotational motions of nanorods provide new insights into complex membrane processes.

In another work, stretchable hydrogels were prepared using gold nanorods.[15] Double-network hydrogels are more similar to rubbers and soft tissues than classic hydrogels. Plasmonic gold nanorods are incorporated into a stretchable double-network hydrogel, composed of alginate and acrylamide. The nanorod-loaded hydrogels could be reversibly stretched, leading to nanorod reversible alignment along the stretch direction as judged by polarized optical spectroscopy. With the proper surface chemistry, hydrogel nanorod composites can be stretched to more than 3000% of their initial length without fracturing. These results show that plasmonic gold nanorods can be well dispersed in multi-component polymer systems, certain surface chemistries can enhance the bulk mechanical properties, and nanorod orientation can be controlled through varying strains on the matrix.

Inhibiting the Aβ aggregation and understanding the kinetic process of Aβ fibrillation are very important for exploring the possible treatment of neurodegenerative diseases. Gold nanorods are shown to inhibit the process of Aβ fibrillation.[16] The results indicate that CTAB-stabilized gold nanorods can efficiently suppress Aβ fibrillation and the inhibition efficiency of larger nanorods is better.

10.7 CONCLUSION

In summary, gold nanorod has been exploited as a drug delivery carrier with the option for co-delivery of multiple drugs and in remote controlled drug delivery. In addition, the light-responsive property is used for *in vivo* targeting and the anisotropic property is used to understand cellular mechanics and in preparation of stretchable hydrogel. We expect that more unique applications will be explored in the near future using the anisotropic property of gold nanorods.

REFERENCES

1. Zheng, J., Cheng, X., Zhang, H., Bai, X., Ai, R., Shao, L. and Wang, J. 2021. Gold nanorods: The most versatile plasmonic nanoparticles. *Chemical Reviews*, 121, 13342–13453.

2. Huang, X., Neretina, S. and El-Sayed, M. A. 2009. Gold nanorods: From synthesis and properties to biological and biomedical applications. *Advanced Materials*, 21, 4880–4910.
3. Dasgupta, S., Auth, T. and Gompper, G. 2014. Shape and orientation matter for the cellular uptake of nonspherical particles, *Nano Letters*, 14, 687–693.
4. Min, Y., Mao, C., Xu, D., Wang J. and Liu, Y. 2010. Gold nanorods for platinum based prodrug delivery. *Chemical Communications*, 46, 8424–8426.
5. Ali, H. R., Ali, M. R. K., Wu, Y., Selim, S. A., Abdelaal, H. F. M., Nasr, E. A. and El-Sayed, M. A. 2016. Gold nanorods as drug delivery vehicles for rifampicin greatly improve the efficacy of combating mycobacterium tuberculosis with good biocompatibility with the host cells. *Bioconjugate Chemistry*, 27, 2486–2492.
6. Chakraborty, D., Tripathi, S., Ethiraj, K. R., Chandrasekaran, N. and Mukherjee, A. 2018. Human serum albumin corona on functionalized gold nanorods modulates doxorubicin loading and release. *New Journal of Chemistry*, 42, 16555–16563.
7. Huang, X., Li, M., Xu, Y., Zhang, J., Meng, X., An, X., Sun, L., Guo, L., Shan, X., Ge, J., Chen, J., Luo, Y., Wu, H., Zhang, Y., Jiang, Q. and Ning, X. 2019. Novel gold nanorod-based HR1 peptide inhibitor for middle east respiratory syndrome coronavirus. *ACS Applied Materials and Interfaces*, 11, 19799–19807.
8. Chen, Y., Chen, X., Wu, D., Xin, H., Chen, D., Li, D., Pan, H., Zhou, C. and Ping, Y. 2021. Delivery of CRISPR/Cas9 plasmids by cationic gold nanorods: Impact of the aspect ratio on genome editing and treatment of hepatic fibrosis. *Chemistry of Materials*, 33, 81–91.
9. Wijaya, A., Schaffer, S. B., Pallares, I. G. and Hamad-Schifferli, K. 2009. Selective release of multiple DNA oligonucleotides from gold nanorods. *ACS Nano*, 3, 80–86.
10. Xiao, Y., Jaskula-Sztul, R., Javadi, A., Xu, W., Eide, J., Dammalapati, A., Kunnimalaiyaan, M., Chen, H. and Gong, S. 2012. Co-delivery of doxorubicin and siRNA using octreotide conjugated gold nanorods for targeted neuroendocrine cancer therapy. *Nanoscale*, 4, 7185–7193.
11. Kolemen, S., Ozdemir, T., Lee, D., Kim, G. M., Karatas, T., Yoon, J. and Akkaya, E. U. 2016. Remote-controlled release of singlet oxygen by the plasmonic heating of endoperoxide-modified gold nanorods: Towards a paradigm change in photodynamic therapy. *Angewandte Chemie International Edition*, 55, 3606–3610.
12. Shiotani, A., Akiyama, Y., Kawano, T., Niidome, Y., Takeshi Mori, T., Katayama, Y. and Niidome, T. 2010. Active accumulation of gold nanorods in tumor in response to near-infrared laser irradiation. *Bioconjugate Chemistry*, 21, 2049–2054.
13. Pissuwan, D., Nose, K., Kurihara, R., Kaneko, K., Tahara, Y., Kamiya, N., Goto, M., Katayama, Y. and Niidome, T. 2011. A solid-in-oil dispersion of gold nanorods can enhance transdermal protein delivery and skin vaccination. *Small*, 7, 215–220.
14. Gu, Y., Sun, W., Wang, G., Zimmermann, M. T., Jernigan, R. L. and Fang, N. 2013. Revealing rotational modes of functionalized gold nanorods on live cell membranes. *Small*, 9, 785–792.

15. Turner, J. G., Og, J. H. and Murphy, C. J. 2020. Gold nanorod impact on mechanical properties of stretchable hydrogels. *Soft Matter*, 16, 6582–6590.
16. Liua, Y., Hea, G., Zhangc, Z., Yina, H., Liua, H., Chena, J., Zhanga, S., Yanga, B., Xub, L.-P. and Zhang, X. 2019. Size-effect of gold nanorods on modulating the kinetic process of amyloid-β aggregation. *Chemical Physics Letters*, 734, 136702.

Toxicology of Gold Nanorod

11

11.1 INTRODUCTION

Gold nanorods are used in wide variety of therapeutic and diagnostic applications.[1–4] One important aspect of all these application is that nanorods should be nontoxic and should not induce any adverse effect to the biological system studied. The toxicity of gold nanorods can arise either due to chemicals associated with nanorod composition or due to the physical processes involved (see Figure 11.1). In terms of chemical composition, gold nanorod is coated with a surfactant (cetyltrimethyl ammonium bromide (CTAB)) double layer. In addition different chemicals are used to cover/coat nanorod surfaces for different applications. They include the exchange of CTAB by thiolated small molecules or polymers, binding of anionic molecules/polymers with cationic CTAB at the surface of the nanorod via electrostatic interaction, encapsulation of nanorods inside polymer micelle and replacement of CTAB layer by silica shell or polymer multilayer.[2] These surface modifications change the behavior of nanorods towards biological environments that include interaction with cell, cellular entry, blood circulation, and trafficking inside the body. As a result, the toxicological aspect will change depending on surface chemistry.

The toxicological effect of gold also depends on different physical environment.[1,3,4] In particular, colloidal nanorods acts as photocatalysts via plasmonic excitation. Under this condition various electrochemical and redox reactions can occur at the nanorod surface that can generate reactive oxygen species. This can enhance the toxicity of nanorods under light exposure. In addition, light absorption of nanorods can induce local heating that can lead to thermal damage of biological components. Moreover, nanorods are known to act as carriers for small molecules. This aspect can enhance the cell delivery of small molecule drugs and induce different toxicological behavior. Here we will discuss all of these chemical and physical aspects of the toxicological effect of gold nanorods, along with selected examples.

DOI: 10.1201/9781003245339-11

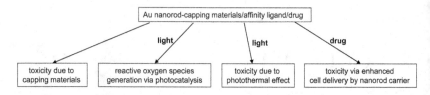

FIGURE 11.1   Schematic representation of the origin of toxicity of gold nanorod either due to chemical composition or due to physical processes.

11.2 TOXICITY OF GOLD NANOROD DUE TO CAPPING MATERIALS

One primary reason for the toxicity of gold nanorods comes from the capping materials. Such materials include cetyltrimethylammonium bromide (CTAB) that is used during nanorod synthesis, coating polymers that are used to exchange CTAB from nanorod surface, and physically/chemically adsorbed molecules at the nanorod surface.[2] The toxicity arises due to the intrinsic toxicity behavior of these coating materials. As these coating materials are weakly bound to nanorod surfaces, they can leach into the biological environment and induce a toxic effect.

Among all the capping materials CTAB is one of the most common capping materials that induce toxicity. Gold nanorods are typically prepared in the presence of a cationic CTAB surfactant and the synthetic process requires a high concentration of CTAB (typically above 100 mM). CTAB is known for membrane damage that induces cytotoxicity. Thus removal of CTAB is necessary for minimizing the cytotoxicity and membrane-compromising properties. However, CTAB capping is important for the colloidal stability of nanorods. Thus, a wide variety of approaches have been reported to replace CTAB with other capping agents and to retain colloidal stability. In addition, the mechanism of cytotoxicity by CTAB-capped nanorod and other coatings are studied. Here we will briefly discuss them.

In one example, CTAB-stabilized nanorods are partially purified by treatment with polystyrenesulfonate, with minimal loss of dispersion stability. However, *in vitro* cytotoxicity assays of polystyrenesulfonate-coated nanorod revealed toxicity in the low to sub-micromolar range, and the source of toxicity is associated with the polystyrenesulfonate–CTAB complex.[5] Further exchange of CTAB-laden polystyrenesulfonate with fresh polyelectrolyte greatly improves biocompatibility due to near complete removal of CTAB. However, polystyrenesulfonate is not effective by itself at stabilizing nanorods in CTAB-depleted suspensions and it should be replaced by more robust coatings for long-term stability under physiological conditions.

Toxicological assays of CTAB-capped nanorod solutions with human colon carcinoma cells reveal that the apparent cytotoxicity is caused by free CTAB in the solution.[6] Overcoating the nanorods with polymers substantially reduces cytotoxicity. The number of nanorods taken up per cell for the different surface coatings is quantitated; the number of nanorods per cell varies from 50 to 2300, depending on the surface chemistry. Serum proteins from the biological media, most likely bovine serum albumin, adsorb to gold nanorods, leading to all nanorod samples bearing the same effective charge, regardless of the initial nanorod surface charge. Such processes should be taken into consideration when examining the biological properties or environmental impact of nanoparticles.

For *in vivo* application, gold nanorods are generally administered via direct injection into the circulation. Thus it is necessary to evaluate their potential adverse effects on blood vessels. In one example, nanorods with various surface modifications were used to evaluate the toxicity and cellular uptake of nanorods into vascular endothelial and smooth muscle cells of isolated rat aortic rings.[7] Surfactant-cappings were replaced and coated with polyelectrolyte or thiolated polyethylene glycol. Cytotoxicity assay showed that thiolated polyethylene glycol-coated nanorods are less toxic than polyelectrolyte-coated nanorods. The difference in toxicity and cellular uptake is linked to free surfactant molecules and protein adsorption, respectively.

In another example, gold nanorods are conjugated with a peptide that recognizes toxic β-amyloid aggregates present in Alzheimer's disease and shows no effects on cell viability in the SH-SY5Y cell line.[8] Furthermore, the irradiation of β amyloid in the presence of the conjugate with near-infrared region irradiation energy reduces the amyloidogenic process, also reducing its cytotoxicity. The chemisorption of the peptide on the surface of gold nanorods increases their stability and reduces their effects on cell viability.

In another example, serum albumin protein corona was coated on CTAB-coated gold nanorods and then the structure of the protein was investigated. The protein adsorption was attributed to at least 12 Au–S bonds and the stable corona reduced the cytotoxicity of CTAB-capped nanorods.[9] This type of protein coating approach improves our understanding of the protective effects of protein coronas against the toxicity of nanomaterials.

The influence of surface chemistry on the interaction between gold nanorods and cell membranes and the subsequent cytotoxicity arising from them in a serum-free cell culture system has been investigated.[10] Results show that nanorod coated with CTAB molecules can generate defects in a cell membrane and induce cell death, mainly due to the unique bilayer structure of CTAB molecules on the surface of the rods rather than their charge.

Compared to CTAB-capped nanorods, positively charged polyelectrolyte-coated (i.e. poly(diallyldimethyl ammonium chloride)) nanorods show improved biocompatibility towards cells. This result indicates that the nature of surface molecules, especially their packing structures on the surface of nanorods rather than surface charges, plays a more crucial role in determining cytotoxicity.

In another work, cellular uptake behavior and cytotoxicity of Au nanorods with various surface coatings were systematically investigated.[11] The surface coating molecules include CTAB, poly(sodium 4-styrenesulfonate), poly(ethylene glycol), mesoporous silica, dense silica, and titanium dioxide. The cellular behavior of nanorods was found to be highly dependent on both the surface coating and the cell type. CTAB, poly(sodium 4-styrenesulfonate), and mesoporous silica-coated nanorods exhibit notable cytotoxicity, while poly(ethylene glycol), dense silica, and titanium dioxide-coated nanorods do not induce cell injury. The cell type plays a preferential role in nanorod cellular uptake. Higher cellular uptake of nanorods was seen in U-87 MG, PC-3, MDA-MB-231, and RAW 264.7 cells, as opposed to HepG2 and HT-29 cells. In addition, nanorod cellular uptake is also highly affected by serum protein binding to the coated surface. This study provides valuable insights into the effects of surface modification on biocompatibility, cellular uptake, as well as biomedical functions of gold nanorods.

In another example, the toxicological effects of two types of polymer-coated gold nanorod on vascular smooth muscle cells are investigated.[12] They include polystyrene sulphonate and poly(diallyldimethyl ammonium chloride). The results show significantly different effects on cells with different surface coatings. It was shown that the binding force and binding probability between nanorods and membranes are closely related to cytotoxicity and cellular responses. All these studies show that coating materials are an important source of gold nanorod toxicity and care must be taken in their synthetic design.

11.3 TOXICITY OF GOLD NANOROD DUE TO PLASMONIC PHOTOCATALYSIS

Plasmonic photocatalysis involves light-induced catalytic reactions at the gold nanorod surface. It is accompanied by three different catalytic processes that include charge transfer-based photoredox catalysis, photo-electrochemical catalysis, and photothermal catalysis.[13] However, photoredox catalysis is the dominant one and this mechanism occurs for gold nanorod-based composite with semiconductor nanoparticles. Under light exposure, the excited surface plasmon promotes redox catalysis via hot electron generation and transfer.[14] For example, the exposure of TiO_2–gold nanorod composite to visible light excites the nanorod plasmon and injects hot electrons to TiO_2 and initiates the catalytic reactions.

Other examples show that plasmonic gold nanorod-based composites can effectively channel the flux of photons into heat to drive catalytic reactions. For example, Pd nanoparticle deposited gold nanorod has been designed for efficient Suzuki coupling reactions under light irradiation.[15] Another work shows that pulsed irradiation on gold nanorods is shown to induce the production of reactive oxygen species (ROS) over a few micrometers in their aqueous environment, particularly hydroxyl radicals.[16] Similarly, Pd-coated Au nanorods are designed where the nanorod acts as a strong plasmonic core, and the Pd shell acts as a nanozyme.[17] The obtained hybrid nanoparticle produced reactive oxygen species via gold nanorod-based photocatalysis and was used to induce cytotoxicity and depolarization of mitochondrial membrane potential. These results show that gold nanorods can induce toxicity under light exposure and this aspect must be taken into account while considering their biomedical application.

11.4 TOXICITY OF GOLD NANOROD DUE TO PHOTOTHERMAL EFFECT

Gold nanoparticle offers a photothermal property that involves light absorption followed by nonradiative energy dissipation in the form of heat.[3] This results in an increase in the temperature of the surrounding medium by tens of degrees. This photothermal effect leads to the damage of the cell membrane, denaturation of intracellular proteins, and induction of cell death via apoptosis. The tunable plasmonic property of gold nanorods from a whole visible to NIR range allows this photothermal effect in various physical environments. This effect has been used for cell therapy applications.

For example, photothermal injury of KB cells is conducted where gold nanorods are localized at the perinuclear region. Cell death has been associated with membrane cavitation and disruption of actin filaments with resultant membrane blebbing.[18] The blebbing results in the disruption of the connections between the cell membrane and cytoskeleton followed by cell death via apoptosis. In another work, monitoring of cancer cell morphology has been investigated during gold-nanorod based photothermal heating.[19] Cancer cells are loaded with gold nanorods at cellular endosomes/lysosomes via endocytic uptake and then irradiated with a pulsed laser. This leads to an explosion of lysosomes and the formation of micron size cavities that led to a rupture of the plasma membrane. Thus the toxic effect of gold nanorods can occur in the presence of light via a photothermal effect.

11.5 TOXICITY OF GOLD NANOROD DUE TO EFFICIENT DELIVERY CARRIER

One of the unique applications of gold nanorods is to act as a delivery carrier for small chemicals/biochemicals. While most of the small chemicals/biochemicals cannot enter into the cell due to the membrane barrier, gold nanorods can allow their entry inside. This is because the larger size of nanorods allows their entry via endocytosis. In particular, small chemicals/biochemicals are either adsorbed at the nanorod surface or electrostatically bound to the coating material of the nanorod and then enter via endocytosis. In other words, gold nanorod offers entry of small chemicals/biochemicals into the cell that can induce different consequences. In particular, known drug molecules, surfactants, and cationic molecules can enter into the cell by this approach and induce more toxic effects that are otherwise not possible without nanorods.

For example, PEGylated gold nanorods are loaded with Pt(IV) prodrug and used as a delivery carrier with superior cytotoxicity against different types of cancer cells.[20] Similarly, gold nanorod is conjugated to rifampicin and is released after uptake into macrophage cells.[21] Nanorods without rifampicin conjugation do not exhibit antimicrobial activity. In another example, lipoic acid appended calix[4]arene (L) has been synthesized where lipoic acid functionality helps L to anchor onto CTAB-capped gold nanorods.[22] As the conjugate of calix[4]arene acts as a host due to the presence of its arene cavity, pyridinium containing guests have been explored to study their complexation, since the pyridinium containing molecules will accumulate in mitochondria in cells. The presence of pyridinium functionality on the guest leads to targeted delivery to mitochondria of nanocomposites as shown by confocal laser scanning microscopy imaging. The nanocomplex has been studied for cancer cell imaging and laser-induced cell killing with the help of plasmonic gold nanorods. When excited at the longitudinal plasmon band of the nanorod, cancer cell killing has been observed due to laser-induced cell death.

11.6 CONCLUSION

In summary, the toxicological aspect of gold nanorods has been briefly reviewed. There are chemical and physical aspects of the toxic effect of gold nanorods. The chemical aspect is linked to the chemical composition that primarily include capping/coating materials and functional materials. The physical aspect is associated with physical conditions such as light exposure or physical properties such as colloidal nanocarrier effect and these conditions can induce plasmonic photocatalysis, photothermal effect, and drug delivery. We expect that the discussed toxicological aspects will help researchers better design nanorod-based therapeutic materials.

REFERENCES

1. Huang, X., Neretina, S. and El-Sayed, M. A. 2009. Gold nanorods: From synthesis and properties to biological and biomedical applications. *Advanced Materials*, 21, 4880–4910.
2. Burrows, N. D., Lin, W., Hinman, J. G., Dennison, J. M., Vartanian, A. M., Abadeer, N. S., Grzincic, E. M., Jacob, L. M., Li, J. and Murphy, C. J. 2016. Surface chemistry of gold nanorods. *Langmuir*, 32, 9905–9921.
3. Abadeer, N. S. and Murphy, C. J. 2016. Recent progress in cancer thermal therapy using gold nanoparticles. *The Journal of Physical Chemistry C*, 120, 4691–4716.
4. Zheng, J., Cheng, X., Zhang, H., Bai, X., Ai, R., Shao, L. and Wang, J. 2021. Gold nanorods: The most versatile plasmonic nanoparticles. *Chemical Reviews*, 121, 13342–13453.
5. Leonov, A. P., Zheng, J., Clogston, J. D., Stern, S. T., Patri, A. K. and Wei, A. 2008. Detoxification of gold nanorods by treatment with polystyrenesulfonate. *ACS Nano*, 2, 2481–2488.
6. Alkilany, A. M., Nagaria, P. K., Hexel, C. R., Shaw, T. J., Murphy, C. J. and Wyatt, M. D. 2009. Cellular uptake and cytotoxicity of gold nanorods: Molecular origin of cytotoxicity and surface effects. *Small*, 5, 701–708.
7. Alkilany, A. M., Shatanawi, A., Kurtz, T., Caldwell, R. B. and Caldwell, R. W. 2012. Toxicity and cellular uptake of gold nanorods in vascular endothelium and smooth muscles of isolated rat blood vessel: Importance of surface modification. *Small*, 8, 1270–1278.
8. Adura, C., Guerrero, S., Salas, E., Medel, L., Riveros, A., Mena, J., Arbiol, J., Albericio, F., Giralt, E. and Kogan, M. J. 2013. Stable conjugates of peptides with gold nanorods for biomedical applications with reduced effects on cell viability. *ACS Applied Materials and Interfaces*, 5, 4076–4085.

9. Wang, L., Li, J., Pan, J., Jiang, X., Ji, Y., Li, Y., Qu, Y., Zhao, Y., Wu, X. and Chen, C. 2013. Revealing the binding structure of the protein corona on gold nanorods using synchrotron radiation-based techniques: Understanding the reduced damage in cell membranes. *Journal of the American Chemical Society*, 135, 17359–17368.

10. Wang, L., Jiang, X., Ji, Y., Bai, R., Zhao, Y., Wu, X. and Chen, C. 2013. Surface chemistry of gold nanorods: Origin of cell membrane damage and cytotoxicity. *Nanoscale*, 5, 8384–8391.

11. Zhu, X.-M., Fang, C., Jia, H., Huang, Y., Cheng, C. H. K., Ko, C.-H., Chen, Z., Wang, J. and Wang, Y.-X. J. 2014. Cellular uptake behaviour, photothermal therapy performance, and cytotoxicity of gold nanorods with various coatings. *Nanoscale*, 6, 11462–11472.

12. Sun, Q., Shi, X., Feng, J., Zhang, Q., Ao, Z., Ji, Y., Wu, X., Liu, D. and Han, D. 2018. Cytotoxicity and cellular responses of gold nanorods to smooth muscle cells dependent on surface chemistry coupled action. *Small*, 14, 1803715.

13. Hana, C., Qi, M.-Y., Tanga, Z.-R., Gong, J. and Xu, Y.-J. 2019. Gold nanorods-based hybrids with tailored structures for photoredox catalysis: Fundamental science, materials design and applications. *Nano Today*, 27, 48–72.

14. Zhang, P., Wang, T. and Gong, J. 2015. Mechanistic understanding of the plasmonic enhancement for solar water splitting. *Advanced Materials*, 27, 5328–5342.

15. Wang, F., Li, C., Chen, H., Jiang, R., Sun, L.-D., Li, Q., Wang, J., Yu, J. C. and Yan, C.-H. 2013. Plasmonic harvesting of light energy for Suzuki coupling reactions. *Journal of the American Chemical Society*, 135, 5588–5601.

16. Labouret, T., Audibert, J.-F., Pansu, R. B. and Palpant, B. 2015. Plasmon-assisted production of reactive oxygen species by single gold nanorods. *Small*, 11, 4475–4479.

17. Fan, H., Li, Y., Liu, J., Cai, R., Gao, X., Zhang, H., Ji, Y., Nie, G. and Wu, X. 2019. Plasmon-enhanced oxidase-like activity and cellular effect of Pd coated gold nanorods. *ACS Applied Materials and Interfaces*, 11, 45416–45426.

18. Tong, L., Zhao, Y., Huff, T. B., Hansen, M. N., Wei, A. and Cheng, J.-X. 2007. Gold nanorods mediate tumor cell death by compromising membrane integrity. *Advanced Materials*, 19, 3136–3141.

19. Chen, C.-L., Kuo, L.-R., Chang, C.-L., Hwu, Y.-K., Huang, C.- K., Lee, S.-Y., Chen, K., Lin, S.-J., Huang, J.-D. and Chen, Y.-Y. 2010. In situ real-time investigation of cancer cell photothermolysis mediated by excited gold nanorod surface plasmons. *Biomaterials*, 31, 4104–4112.

20. Min, Y., Mao, C., Xu, D., Wang J. and Liu, Y. 2010. Gold nanorods for platinum based prodrug delivery. *Chemical Communications*, 46, 8424–8426.

21. Ali, H. R., Ali, M. R. K., Wu, Y., Selim, S. A., Abdelaal, H. F. M., Nasr, E. A. and El-Sayed, M. A. 2016. Gold nanorods as drug delivery vehicles for rifampicin greatly improve the efficacy of combating mycobacterium tuberculosis with good biocompatibility with the host cells. *Bioconjugate Chemistry*, 27, 2486–2492.

22. Nag, R., Kandi, R. and Rao, C. P. 2018. Host–guest complexation of a lipoic acid conjugate of calix[4]arene with pyridinium moiety on gold nanorods for mitochondrial tracking followed by cytotoxicity in HeLa cells under 633 nm laser light. *ACS Sustainable Chemistry and Engineering*, 6, 8882–8890.

Advantages of Gold Nanorod 12

12.1 INTRODUCTION

Advancements in the field of gold nanorods have shown that gold nanorods have some unique properties that make them different from spherical and other anisotropic gold nanoparticles. Coupled with the commercial availability of gold nanorods and the availability of a large variety of surface chemistry/modification, application potentials are widely open. In this chapter, we describe the unique aspects of gold nanorods along with their application potential. Their unique properties include length-dependent optical property, anisotropic functionalization, anisotropic chemical reactivity, and length-dependent self-assembly property. In addition, there are some unique applications of gold nanorods for single molecule detection via enhanced fluorescence property or surface-enhanced Raman signal, understanding the membrane property via rotational modes of gold nanorods, and fabrication of metamaterials with the tunable optical property.

Although there are several well-known anisotropic nanoparticles such as gold-based nanoparticles (e.g. platelets, triangular nanoprisms, nanocubes, nanostars, and nanoshells), silver-based anisotropic nanoparticles (e.g. nanorods, platelets, nanoprisms, and nanostars), and other nanoparticles (e.g. CdSe nanorods, TiO_2 nanorods, hydroxyapatite nanorods/nanowires, and $BaTiO_3$ nanorods), the above mentioned specific properties and related applications are unique to gold nanorods. Here we will summarize them with some selected examples.

12.2 LENGTH-DEPENDENT OPTICAL PROPERTY

Gold nanorods have a unique advantage over spherical or other anisotropic shapes in terms of their optical property. Their colloidal solutions have a length-dependent plasmonic property that is tunable in the range of 500–1200 nm.[1,2] In contrast, the plasmonic property of gold spheres can be tuned only from 500 to 600

DOI: 10.1201/9781003245339-12

nm by varying their size. In addition, plasmonic tunability is either restricted to narrow regions or plasmon bands are relatively broad for other shapes.[3,4] Although silver-based anisotropic nanoparticles (e.g. nanorods, platelets, nanoprisms, and nanostars) are good competitors for gold nanorods, they are less biocompatible and more susceptible to shape change under adverse conditions.[5,6]

This tunable plasmonic property has the advantage for gold nanorods as this is sensitive to surface chemical environment and this property is used for various optical detection applications. Moreover, this property has an advantage as a surface-enhanced Raman spectroscopy (SERS) substrate, as a wide variety of lasers in the range of 500–1200 nm can be used for their plasmonic excitation and SERS detection.

12.3 ANISOTROPIC FUNCTIONALIZATION AT EITHER END OF NANOROD

One unique property of gold nanorods is that they can be anisotropically functionalized. This is particularly because the capping CTAB surfactant is differentially bound to nanorods. CTAB binds relatively strongly to the side faces but loosely to the tips of nanorods.[7] This feature can be used for anisotropic functionalization of nanorods followed by different applications.[8–10] More specifically, small molecule thiols and thiolated polymers are attached at nanorod tips without attaching to the side face of the nanorods. Next, such functional nanorods are used for end-to-end linear assembly of nanorods or directly used for plasmon-based analyte detection.[8,11–13] As either end of the nanorods are the plasmonic hot spots, such functionalization of nanorods are used in SERS-based detection applications.[14]

Alternatively, gold nanorods with porous TiO_2 caps at either end have been synthesized.[15,16] The positioning of photoactive species at either end of gold nanorods has the advantage of utilizing the electric field enhancement effect at those ends. These structures are used for plasmon-enhanced reactive oxygen species (ROS) generation and hydrogen generation from water.[15,16]

12.4 ANISOTROPIC CHEMICAL REACTIVITY AT EITHER END OF NANOROD

Another unique aspect of gold nanorods is their anisotropic chemical reactivity. In particular, oxidative etching/dissolution of nanorods can occur along the side faces of nanorods, particularly due to less effective CTAB capping and less stable atomic planes of gold. Such anisotropic etching of gold nanorods induces a decrease in the nanorod aspect ratio and leads to the blue shifting of the longitudinal plasmon band. This aspect has been used for the optical detection of various oxidizing agents such as metal ions,[17] persulfates,[18] and reactive oxygen species or enzymes that generate reactive oxygen species.[19]

For example, thiocyanate functionalized colloidal gold nanorod is subjected to hydrogen peroxide and Co^{2+} ions where Co^{2+} ions trigger the Fenton-like reaction and generate superoxide radical. As a result, the gold nanorods are gradually etched preferentially along the longitudinal direction. This is accompanied by a color change from green to red that is used for Co^{2+} ion detection in nanomolar concentration.[17]

12.5 LENGTH-DEPENDENT SELF-ASSEMBLY

Another unique aspect of gold nanorods is their length-dependent self-assembly property. Nanorods self-assemble into different types of liquid crystalline structures at high concentrations, depending on their length.[20–22] The smaller nanorods usually assemble into smectic ordering, longer nanorods try to assemble into columnar or ribbon-like structures, and nanowires make bundle structures. Such self-assembly structures are driven by entropy as well as capped surfactants.

Such self-assembly properties have been utilized for making 2D and 3D assembly of gold nanorods via Langmuir–Blodgett films, drying-induced assembly, and in other approaches.[23] These assembled structures are used for generating plasmonic hot spots for surface-enhanced Raman-based detection applications and device fabrication for solar energy conversion applications.[23] In addition, these assembly properties are used for shape separation.[20, 21]

12.6 OTHER PROPERTIES

In addition to the described unique aspects of gold nanorods, there are some specific applications that use the shape-dependent specific properties of gold nanorods. They include the enhanced fluorescence/surface-enhanced Raman property for single molecule detection, length-dependent local heat generation property, understanding the membrane property via rotational modes of gold nanorods, and fabrication of optical metamaterials for chiral sensing.

For example, gold nanorods with length to width ratio of 2.0–5.4 show a million times enhanced fluorescence.[24] While nanospheres of 35 nm diameter did not emit light, nanorods are found to have a fluorescence of quantum efficiency which is 6–7 orders of magnitude higher than that of the metal. This emission occurs due to electron and hole interband recombination. The increase in the emission yield results from the enhancement effect of the incoming and outgoing electric fields via coupling to the surface plasmon resonance in the rods. This enhancement is similar to the fluorescence and the Raman enhancement on the rough surfaces of noble metals. Similarly, individual gold nanorod is used to enhance the fluorescence quantum yield of methylene blue for single-molecule detection via redox-induced fluorescence blinking dynamics.[25]

In plasmonic photothermal therapy, the relative contributions of absorption vs scattering properties are critical. Gold nanorods with different aspect ratios can be used to modulate photothermal properties. It is shown that larger gold nanorods with high scattering still possess strong photothermal capabilities, which rival that of smaller rods on an ensemble level, and surpass small rods in both single-particle temperature increases and volume-normalized extinction.[26] In another example, an effective strategy for the directional assembly of anisotropic nanoparticles is used for the assembly of gold spheroids and nanorods.[27] The well-defined secondary plasmonic hot spots between the coupled gold spheroids exhibit the capability for single molecule detection.

The unique capabilities of visualizing and understanding rotational motions of functional gold nanorods on live cell membranes provide new insights into complex membrane processes.[28] Gold nanorods with functionalized surfaces that interact with the membrane, exhibit distinct rotational modes that allow the observation of early events of membrane properties. The single particle orientation and rotational tracking technique are combined here with correlation analysis to identify the fundamental rotational modes of single gold nanorod probes in live cell imaging experiments.

Gold nanorod has been used for the fabrication of tunable optical metamaterials by manipulating the directional self-assembly via programmable surface adapters.[29] A DNA surface adapter is used that can programmably self-assemble into various chiral supramolecular architectures, thereby regulating the chiral directional "bonding" of gold nanorods. Distinct optical chirality relevant to the ensemble conformation is demonstrated from the assembled novel stair-like and coil-like gold nanorod chiral metastructures, which is strongly affected by the spatial arrangement of neighboring nanorod pair.

In addition, length-dependent change of cell uptake can be expected for gold nanorods that has potential biomedical applications. It has been proposed that endocytic uptake of rod-like nanoparticles will depend on nanorod length.

It is shown theoretically that longer nanorods enter via a submarine mode (i.e. side first with their long edge parallel to the membrane) but shorter nanorods enter via a rocket mode (i.e. tip first with their long edge perpendicular to the membrane).[30] However, such uptake has not yet been experimentally demonstrated.

12.7 CONCLUSION

The most unique aspects of gold nanorods and their application potentials are discussed in this chapter. These include length-dependent optical property and self-assembly property, functionalization and reactivity along either end of nanorods, single-molecule detection via enhanced fluorescence and surface-enhanced Raman signal, and investigating cell membrane dynamics via nanorod rotational dynamics and fabrication of metamaterials. We expect that these unique properties will be utilized further for the development of this field.

REFERENCES

1. Huang, X., Neretina, S. and El-Sayed, M. A. 2009. Gold nanorods: From synthesis and properties to biological and biomedical applications. *Advanced Materials*, 21, 4880–4910.
2. Chen, H., Shao, L., Lia, Q. and Wang, J. 2013. Gold nanorods and their plasmonic properties. *Chemical Society Reviews*, 42, 2679–2724.
3. Sau, T. K., Rogach, A. L., Döblinger, M. and Feldmann, J. 2011. One-step high-yield aqueous synthesis of size-tunable multispiked gold nanoparticles. *Small*, 7, 2188–2194.
4. Brinson, B. E., Lassiter, J. B., Levin, C. S., Bardhan, R., Mirin, N. and Halas, N. J. 2008. Nanoshells made easy: Improving Au layer growth on nanoparticle surfaces. *Langmuir*, 24, 14166–14171.
5. Jana, N. R. and Pal, T. 2007. Anisotropic metal nanoparticles for use as surface-enhanced Raman substrates. *Advanced Materials*, 19, 1761–1765.
6. Zhang, O., Li, N., Goebl, J., Lu, Z. and Yin, Y. 2011. A systematic study of the synthesis of silver nanoplates: Is citrate a "magic" reagent? *Journal of the American Chemical Society*, 133, 18931–18939.
7. Khelfa, A., Meng, J., Byun, C., Wang, G., Nelayah, J., Ricolleau, C., Amara, H., Guesmi, H. and Alloyeau, D. 2020. Selective shortening of gold nanorods: When surface functionalization dictates the reactivity of nanostructures. *Nanoscale*, 12, 22658–22667.

8. Sudeep, P. K., Joseph, S. T. S and Thomas, K. G. 2005. Selective detection of cysteine and glutathione using gold nanorods. *Journal of the American Chemical Society*, 127, 6516–6517.

9. Stewart, A. F., Lee, A., Ahmed, A., Ip, S., Kumacheva, E., and Walker, G. C. 2014. Rational design for the controlled aggregation of gold nanorods via phospholipid encapsulation for enhanced Raman scattering. *ACS Nano*, 8, 5462–5467.

10. Yilmaz, H., Bae, S. H., Cao, S., Wang, Z., Raman, B. and Singamaneni, S. 2019. Gold-nanorod-based plasmonic nose for analysis of chemical mixtures. *ACS Applied Nano Materials*, 2, 3897–3905.

11. Yu, C. and Irudayara, J. 2007. Multiplex biosensor using gold nanorods. *Analytical Chemistry*, 79, 572–579.

12. Zhao, Y., Tong, L., Li, Y., Pan, H., Zhang, W., Guan, M., Li, W., Chen, Y., Li, Q., Li, Z., Wang, H., Yu, X.-F. and Chu, P. K. 2016. Lactose-functionalized gold nanorods for sensitive and rapid serological diagnosis of cancer. *ACS Applied Materials and Interfaces*, 8, 5813–5820.

13. Beiderman, M., Ashkenazy, A., Segal, E., Barnoy, E. A., Motiei, M., Sadan, T., Salomon, A., Rahimipour, A., Fixler, D. and Popovtzer, R. 2020. Gold nanorod-based bio-barcode sensor array for enzymatic detection in biomedical applications. *ACS Applied Nano Materials*, 3, 8414–8423.

14. Pardehkhorram, R., Alshawawreh, F. A., Goncales, V. R., Lee, N. A., Tilley, R. D. and Gooding, J. J. 2021. Functionalized gold nanorod probes: A sophisticated design of SERS immunoassay for biodetection in complex media. *Analytical Chemistry*, 93, 12954–12965.

15. Wu, B., Liu, D., Mubeen, S., Chuong, T. T., Moskovits, M. and Stucky, G. D. 2016. Anisotropic growth of TiO_2 onto gold nanorods for plasmon-enhanced hydrogen production from water reduction. *Journal of the American Chemical Society*, 138, 1114–1117.

16. He, L., Mao, C., Brasino, M., Harguindey, A., Park, W., Goodwin, A. P. and Cha, J. N. 2018. TiO_2-capped gold nanorods for plasmon-enhanced production of reactive oxygen species and photothermal delivery of chemotherapeutic agents. *ACS Applied Materials & Interfaces*, 10, 27965–27971.

17. Zhang, Z., Chen, Z., Pan, D. and Chen, L. 2015. Fenton-like reaction-mediated etching of gold nanorods for visual detection of Co^{2+}. *Langmuir*, 31, 643–650.

18. Jana, N. R., Gearheart, L., Obare, S. O. and Murphy, C. J. 2002. Anisotropic chemical reactivity of gold spheroids and nanorods. *Langmuir*, 18, 3, 922–927.

19. Saa, L., Coronado-Puchau, M., Pavlov, V. and Liz-Marzan, L. M. 2014. Enzymatic etching of gold nanorods by horseradish peroxidase and application to blood glucose detection. *Nanoscale*, 6, 7405–7409.

20. Jana, N. R. 2004. Shape effect in nanoparticle self-assembly. *Angewandte Chemie International Edition*. 43, 1536–1540.

21. Jana, N. R. 2003. Nanorod shape separation using surfactant assisted self-assembly. *Chemicals Communications*, 1950–1951.

22. Vaia, R. A. 2010. Depletion-induced shape and size selection of gold nanoparticles. *Nano Letters*, 10, 1433–1439.

23. Scarabelli, L., Hamon, C. and Liz-Marzan, L. M. 2017. Design and fabrication of plasmonic nanomaterials based on gold nanorod supercrystals. *Chemistry of Materials*, 29, 15–25.

24. Mohamed, M. B., Volkov, V., Link, S. and El-Sayed, M. A. 2000. The 'lightning' gold nanorods: Fluorescence enhancement of over a million compared to the gold metal. *Chemical Physics Letters*, 317, 517–523.
25. Zhang, W., Caldarola, M., Pradhan, B. and Orrit, M. 2017. Gold nanorod enhanced fluorescence enables single-molecule electrochemistry of methylene blue. *Angewandte Chemie International Edition*, 2017, 56, 3566–3569.
26. Meyer, S. M., Pettine, J., Nesbitt, D. J. and Murphy, C. J. 2021. Size effects in gold nanorod light-to-heat conversion under femtosecond illumination. *The Journal of Physical Chemistry C*, 125, 16268–16278.
27. Wang, G., Akiyama, Y., Takarada, T. and Maeda, M. 2016. Rapid non-crosslinking aggregation of DNA-functionalized gold nanorods and nanotriangles for colorimetric single-nucleotide discrimination. *Chemistry – A European Journal*, 22, 258–263.
28. Gu, Y., Sun, W., Wang, G., Zimmermann, M. T., Jernigan, R. L. and Fang, N. 2013. Revealing rotational modes of functionalized gold nanorods on live cell membranes. *Small*, 9, 785–792.
29. Lan, X., Su, Z., Zhou, Y., Meyer, T., Ke, Y., Wang, Q., Chiu, W., Liu, N., Zou, S., Yan, H. and Liu, Y. 2017. Programmable supra-assembly of a DNA surface adapter for tunable chiral directional self-assembly of gold nanorods. *Angewandte Chemie International Edition*, 56, 14632–14636.
30. Dasgupta, S., Auth, T. and Gompper, G. 2014. Shape and orientation matter for the cellular uptake of nonspherical particles. *Nano Letters*, 14, 687–693.

Index

Printed in the United States
by Baker & Taylor Publisher Services